SIMON FRASER UNIVERSITY
W.A.C. BENNETT LIBRARY

Understanding and Communicating
SOCIAL INFORMATICS

Understanding and Communicating
SOCIAL INFORMATICS

A Framework for Studying and Teaching the Human Contexts of Information and Communication Technologies

Rob Kling
Howard Rosenbaum
Steve Sawyer

 **Information Today, Inc.**
Medford, New Jersey

First printing, 2005

Understanding and Communicating Social Informatics: A Framework for Studying and Teaching the Human Contexts of Information and Communication Technologies

This book has been funded in part by the National Science Foundation under Grant No. IRI-9714211. Any opinions, findings, conclusions, or recommendations expressed in this material are those of the authors and do not necessarily reflect the views of the National Science Foundation.

Publisher's Note: The authors and publisher have taken care in preparation of this book but make no expressed or implied warranty of any kind and assume no responsibility for errors or omissions. No liability is assumed for incidental or consequential damages in connection with or arising out of the use of the information or programs contained herein.

Many of the designations used by manufacturers and sellers to distinguish their products are claimed as trademarks. Where those designations appear in this book and Information Today, Inc. was aware of a trademark claim, the designations have been printed with initial capital letters.

Library of Congress Cataloging-in-Publication Data

Kling, Rob.
  Understanding and communicating social informatics : a framework for studying and teaching the human contexts of information and communication technologies / Rob Kling, Howard Rosenbaum, Steve Sawyer.
    p. cm.
  Includes bibliographical references and index.
  ISBN 1-57387-228-8
    1. Computers and civilization. I. Rosenbaum, Howard. II. Sawyer, Steve, 1960– III. Title.
    QA76.9.C66K54 2005
    303.48'33--dc22

                                                                    2005017205

President and CEO: Thomas H. Hogan, Sr.
Editor-in-Chief and Publisher: John B. Bryans
Managing Editor: Amy M. Holmes
VP Graphics and Production: M. Heide Dengler
Book Designer: Kara Mia Jalkowski
Cover Designer: Dana Kruse
Copyeditor: Dorothy Pike
Proofreader: Susanne Bjorner
Indexer: Sharon Hughes

Contents

CHAPTER 1

CHAPTER 2

CHAPTER 3

CHAPTER 6
**Communicating Social Informatics Research to Professional
and Research Communities**

CHAPTER 7

Figures and Tables

Foreword

By William H. Dutton

What would be the impact of a national identification card? Will the Internet democratize diplomacy, or the political process generally? Will people use next-generation mobile phones and with what effect on everyday life?

Popular answers to such questions are offered by many pundits, filling the press with utopian and dystopian forecasts. However, the serious research needed to address questions about the social shaping, use, and societal implications of information and communication technologies (ICT) requires collaborative research across a variety of disciplines. These questions are not the preserve of any one discipline, be it computer science or sociology.

Understanding and Communicating Social Informatics illustrates the value of Social Informatics, defined here as a multi-disciplinary approach to empirical research on the social aspects of ICT. The authors explain how this approach has been applied in a diverse range of areas, and provide rich examples to demonstrate the important insights it has revealed. In addition, this book conveys the craftsmanship of the authors in their careful observation of cases and their insightful interpretation of complex interrelationships among social and technical processes and outcomes.

As you read the pages that follow, you will appreciate the qualities that have made Social Informatics a recognized approach to the study of ICT. The major contribution to the origins and future of this approach made by Rob Kling and his colleagues might be less obvious to many readers, particularly the degree to which Social Informatics is anchored not in any mechanical multi-disciplinarity, but in true intellectual skillfulness. My goal here is to convey this through a brief overview of the origins, establishment, art, and maintenance of Social Informatics.

The Origins of Social Informatics

The book's lead author, Rob Kling, who died on May 15, 2003, was the founder and central figure in conceiving and promoting the idea of Social Informatics as an approach to research on ICT. I collaborated with him from 1974 to 1979 at what was then called the Public Policy Research Organization (PPRO) at the University of California at Irvine.

Earlier, Rob completed his doctorate in computer science at Stanford University, focusing on artificial intelligence. He worked in the Artificial Intelligence Center at the Stanford Research Institute and then at the University of Wisconsin at Madison in 1970 as an assistant professor of computer science, before joining Irvine in 1973. In Irvine he worked with Kenneth Kraemer, a professor of management and gifted manager of research, James Danziger, an eloquent speaker and political scientist, Alexander Mood, a renowned statistician, and others, including Ken's doctoral student, John Leslie King, now Dean of the School of Information at the University of Michigan. Together they developed a successful National Science Foundation proposal for an evaluation of urban information systems (URBIS), which ran from 1974 to 1979.

I was fortunate to join this URBIS Group in 1974, becoming one of the project's principal investigators. URBIS was one of the first large scale evaluations of computers in government, and one of the early multidisciplinary studies of the social impact of computing. With some major exceptions, such as Alan Westin and Michael Baker's (1972) study of computers and privacy, most work on the social and organizational implications of computing was based more on logical extrapolations from the features of computer technology than on close empirical inquiry into the question: How do people actually design, implement, and use technologies in specific social contexts, and with what effects?

In the 1970s, computer scientists were the major contributors to debates about the social impacts of computers. Separately, social scientists, such as Daniel Bell (1974), were focusing on societal trends connected to the development of information technology, leading to new perspectives on the nature of post-industrial, information societies. But these and other more disciplinary streams of research and debate seldom connected.

It was in the midst of these various treatments of computers and society that Kenneth Kraemer put together the URBIS Group to focus on the impact of computers in American local governments. Through its process of conducting a focused empirical research project, the Group began to gain increasing appreciation for the value of multidisciplinary inquiry. Rob immersed himself in sociological theory and research, from symbolic interaction to organizational behavior, while he tutored us on computing technologies. He was the first among us who sought to label the endeavour we were jointly pursuing.

I have vivid memories of discussions from the early 1980s in which Rob was asking me if I thought the term Social Informatics was promising as a phrase that might capture this emerging approach. I thought not. I did not dissuade Rob and with the benefit of hindsight, I was wrong. Rob claims that the origins of Social Informatics stemmed from discussions in 1996 at a workshop on social aspects of digital libraries (Kling, 1999). However, I would maintain that his search for a label—if not the concept of Social Informatics—for this approach to a new field began at least a decade earlier.

By 1996, many other terms had been coined. Some focused on naming the new ICT emerging from the convergence of computing, telecommunications, and management science, such as: Information Technology, new media, "compunications," the French "telematique" and "informatique," and informatics, most often used in Europe. Others focused on capturing the social aspects of computing, such as studies of the "social impact of computing," "social analysis of computing," "computers and society," "computer-mediated communication" studies, and more. The concept of Social Informatics joined together these naming traditions.

Alan Kay (in Frenkel, 1994) wrote that the "best way to predict the future is to invent it." And that is what Rob did. Not persuaded by my views or those of other critics, Rob, his students, and a growing cadre of colleagues and networks of like-interested scholars began to use and promote the concept of Social Informatics. He coined the term, promoted it, developed the Center for Social Informatics in the School of Library and Information Science at Indiana University, and devoted much effort as editor of *The Information Society* to supporting its practice.

Multidisciplinary Craftsmanship

The intellectual craftwork that underpins its multidisciplinary research is critical to the success of Social Informatics, and it was through this craftsmanship that Rob developed and sustained his idea. He was a role model of multidisciplinary research, having become extremely well versed in both computer science and the social sciences. This enabled him to question social scientists on their own ground, avoiding the all too frequent lack of communication across disciplines. More importantly, he had such a creative, analytical mind that he was able to spot and unravel taken-for-granted assumptions about both computing technologies and potential social implications in ways that generated new insights.

For instance, I recall him responding in a panel discussion to a rather doctrinaire assertion that any rules or regulations would undermine the freedom of computer-based communication by moving seamlessly into a description of the rules—signs, limits, laws, and technical constraints—he faced driving his car to work. "Do traffic restrictions limit freedom or enable more people to drive more safely from anywhere to anywhere?" he asked.

In another instance, on the URBIS Project in which I collaborated with Rob and others, we struggled for weeks to characterize computing in most local governments. Rob refused to take on board the conventional terms, such as computer systems, because so much computing in organizations was not well integrated into a "system," such as the computers stored in a back room, or staff who had no idea how to use computer applications. We finally arrived at the concept of the "computer package" to capture the often loose assortment of people, equipment, and techniques that defined computing in local governments (Danziger et al., 1982, pp. 22–50).

This ability to move between concrete, detailed cases to more general theoretical concepts and themes was central to Rob's approach. Similarly, many of the illustrations developed in this book show how the routine can be looked at in creative ways to tease out the complex interrelationships between the technical and the social.

Together, Rob, Howard, and Steve have produced a work that will help sustain social informatics because it graphically sheds light on the degree to which multidisciplinary inquiry is not a step-by-step cookbook approach. At its heart are individual scholars who have the patience, collaborative sensibilities, powers of observation, analytical agility, and intellectual aptitude to look at complex social and technical systems and find general patterns and themes that extend well beyond the particular case.

The Continuity of Social Informatics

Was Rob Kling's approach unique to him? Could he teach the attitudes and skills that would enable others to approach the social analysis of computing in similar ways? It is an open question: Can successful academics transfer their intellectual capital to others? Are they gifted individuals, or do they have a method they can impart through apprenticeship or their teaching? In some fields, such as survey research, the transferability of methods is relatively well established. In some areas of qualitative research, there is often more dependence on the art of individual researchers. For example, it is very difficult for a qualitative researcher, such as an ethnographer, to delegate field research. Without being there, it is difficult to gain the same insights.

Rob not only invented Social Informatics, but through his founding and leadership of the Center for Social Informatics, he also enabled it to move beyond himself through generations of students, the work of the Center, and in his influence on countless others who worked with him as a valued partner or know him only from his lucid writing. Howard and Steve worked for over a year to complete this book after Rob's death. It is a testament to his influence on the field and the respect in which colleagues hold him that Rob remains the first author.

The following pages bring Social Informatics to the reader through general themes and concrete examples. It is a product of decades of research on the social aspects of information and communication technologies, from the mainframe to the Internet and beyond. And it is a tribute to the inspiration and supreme intellectual craftsman of Social Informatics: Rob Kling.

William H. Dutton

Oxford, May 2005

References

Bell, D. (1974). *The coming of post-industrial society: A venture in social forecasting.* London: Heinemann (originally published, New York: Basic Books, 1973).

Danziger, J. N., Dutton, W. H., Kling, R., & Kraemer, K. L. (1982). *Computers and politics.* New York: Columbia University Press.

Frenkel, K. A. (1994). A conversation with Alan Kay. *Interactions, 1*(2), 13–22.

Kling, R. (1999). What is Social Informatics and why does it mattter? *D-Lib Magazine,* 5(1), January. Available at http://www.dlib.org/dlib/january99/kling/01kling.html

Westin, A., & Baker, M. (1972). *Databanks in a free society: Computers, record-keeping, and privacy.* New York: Quadrangle Books.

Acknowledgments

We are particularly pleased to thank John Bryans and Amy Holmes at Information Today, Inc., who worked with us to make this book become a reality. Along the way, the works and work were tremendously strengthened by Sharon Ahlers's critical insights and editorial guidance. We thank Holly Crawford and Susie Weisband for their formative work and comments on early drafts of the book. We thank Sara Reagor for doing the hard knowledge-work of collecting references and citations. In Appendices C and D, we have listed the fifty people who read and commented on the many interim versions of this book; their comments have greatly improved the book. Finally, we appreciate the support of our families: Being consumed by writing a book requires the love and support of others.

Preface

In 1997 the National Science Foundation funded Rob Kling and Howard Rosenbaum to host a workshop at Indiana University in Bloomington, Indiana. The workshop was organized around three objectives, the most important of which was the articulation of a new, transdisciplinary research domain called Social Informatics. A second objective was to outline a set of research issues around which Social Informatics would be centered, and, third, to clarify a research agenda for this domain. Thirty scholars from a wide range of disciplines attended, and during three days of intensive work, accomplished these objectives. The workshop had other outcomes as well, including raising participant's awareness of the commonalities in their work despite their diverse intellectual homes and providing a common identity for those pursuing questions about information and communication technologies, people, and work.

Looking back, we can now see that, led by Rob Kling's energy and vision, the participants' efforts helped to crystallize the Social Informatics movement. Steve Sawyer attended the workshop and joined with Howard and Rob soon after to "write a book about Social Informatics." We initially saw this book as a summary of the 1997 Social Informatics workshop that would provide a record of the collective enthusiasm generated by the participants and the insights that came from this event.

We worked on the book over the next two years with the assistance of Holly Crawford and Suzie Weisband. However, our academic lives intervened: We wrote more directed articles, conducted specific studies, and organized tracks at conferences. Although we set the project aside temporarily, our collective actions helped to push forward Social Informatics. At Rob's urging, we completed the report on the workshop, which we delivered to the National Science Foundation in 2000. In that document, we pointed with pride to the increasing prevalence of Social Informatics as a disciplinary label, and highlighted the increasing levels of understanding and acceptance of Social Informatics work. In that document we also noted the emerging recognition of Social Informatics as a viable field of study in computing, just as has happened to human–computer interaction, computer-supported cooperative work, and software engineering over the past twenty years. We also lamented the lack of a central monograph on Social Informatics.

Rob, Steve, and Howard returned in earnest to the book in 2001. We had left a draft of the book on Rob's Web site, and it was being downloaded with increasing frequency. People were asking when the book would be published and our response was that we were "almost done with it." At this time, John Bryans of Information Today, Inc. helped us to refine the concept; his enthusiasm and support made the book real. With his help we were able to hire Sharon Ahlers to provide editorial assistance. Sharon has a passion for both Social Informatics and excellent writing, and we relied on both her insights and remarkable patience as we worked three people's writing into one manuscript.

Three years have passed since we began reworking this book, reflecting the passion we have for Social Informatics. Each of us championed parts; all of us worked through each word of each chapter. This book is a true collaboration; it is our work and we cannot really identify solely individual contributions. Rob's breadth, depth, vision, command of so many literatures, views on writing, specific audiences whom he sought to influence, and collegiality drove us to work at high levels of quality. And he led by example. Howard's intellectual precision, grasp of material, and care with language helped merge three authors' voices into one. Steve's analytic approach, collaborative nature, unflagging enthusiasm for the project, and project management skills helped to bring the book into print.

We have enjoyed writing this book. It has taken longer than we expected, but we believe that this aging makes the book better. In the pages that follow, we focus on conceptualizing computing and social change, getting away from focusing on particular technologies to identify more abstract principles, common findings, and issues. No longer a summary, in this book we lay out the principles of Social Informatics, bring together and explore the common elements of a Social Informatics perspective, tie this perspective to particular areas of focused research and teaching, and lay out Social Informatics for those outside this area. This book is an overview, designed to be short, readable in pieces, and directed to our colleagues and those who wish to engage with the concepts and issues of computing from a social perspective.

Rob Kling's untimely death in May 2003 left us shaken but resolute. Rob was an advocate, rallying point, intellectual force, and seemingly endless energy source for Social Informatics. His presence is in every page of this book, especially the last chapter, which he left us in draft form. Steve took over the lead on the final chapter and we tip-toed through Rob's words for a bit, until Jonathan Grudin noted that Rob's work was always in play, always undergoing change, and that we'd be wrong to reify words or fix an idea as immutable. We have now finished the work. The book's role is still as a primer—claims of legacy are best left to Rob's writings and, more importantly, his leadership in making Social Informatics. Rob worried that Social Informatics had to be more than him. It is. We are proud and delighted to enter this book into the broader intellectual discourse that Rob so loved.

Howard Rosenbaum and Steve Sawyer

CHAPTER 1

Introduction to Social Informatics

1.1 The Disconnection Between Popular and Scholarly Discussion

What are the effects of computerization on our society? For instance, what does it mean when people say that the Internet will re-shape society? How much will people telecommute and how will telecommuting practices change the way we work? Will the increased use of computers and the presence of the Internet radically reduce the enrollments in colleges and universities whose programs are "place-based"? How will the Web change the ways that people search for and use medical information? These questions are a sample of the ongoing and ever-increasing discussion about the ways in which computer-based systems—more broadly "information and communication technologies (ICTs)"[1]—are playing powerful roles in reshaping organizations and social relations. These discussions take place in a variety of arenas, including personal conversations, newspaper articles, writings by pundits, textbooks about designing and/or managing ICTs, policy analyses, careful professional accounts in professional magazines, and systematic academic research.

Examples of the influence of ICTs on social life are hard to escape. The World Wide Web has become ubiquitous in ads and newspapers, and computing systems are now integral to our banking, transportation, medical, educational, and, increasingly, retail systems in North America. Further, many employers are seeking "computer-literate" workers in a wide variety of occupations, from manufacturing to marketing.

Social Informatics refers to the body of systematic research about the social aspects of ICTs. Unfortunately, the findings and theories from most of the systematic research rarely appear in the popular media or even in many of the textbooks and policy analyses. Rather, the research can usually only be found in books and journals that are available from specialty publishers or certain scientific societies, or are located in specialized research libraries. The interested layperson or professional who goes to a large chain bookstore would have trouble finding these materials. Instead, he or she will more readily find materials written by pundits and journalists whose writings don't seem to reflect the research.

With this book, we introduce you to the systematic, rigorous, and empirically based research that has focused on these and other computerization issues. Further, we provide you with a means to draw on the large and growing body of research that has addressed these questions and to organize the collected findings conceptually.

The public discourse on how ICTs change organizational and social life is being shaped by personal experiences, journalists' booking, pundits' predictions, technologically utopian and dystopian accounts in a range of literatures, and high-level policy discussions. This discourse on ICTs and social change pervades our lives, even though many discussions of the roles of ICTs focus primarily on technical features. Moreover, many of the popular discussions about the roles and socioeconomic effects of ICTs are based on vivid, compelling and well-articulated, but essentially armchair, or anecdotal, speculation.

A middle zone also exists, between the less systematic and a-theoretical popular accounts and the more systematic, empirically grounded and theoretically informed research studies. These are the systematic professional accounts of ICTs and social change that are written by sophisticated and careful journalists (e.g., Garfinkel, 2000) or are the products of careful empirical research by the staffs of public agencies (i.e., National Telecommunications and Information Administration, 1998, 1999).

By "disconnected discussions," we refer to the way that the public discourse on the roles of ICTs in society takes place almost independent of the accumulated body of knowledge that has arisen from careful empirical research. Much of the enthusiasm for dot-coms in the late 1990s was based on the assumption that firms that relied upon the Internet for business must be less expensive to run than their "brick and mortar" counterparts. This assumption ignored research about the "total costs" of organizational information systems, including infrastructure that ranged from relatively costly high-speed networks, software maintenance, and technical training for new staff. In contrast with the best research, it ignored the investments in "complementary resources," such as intensive advertising and warehouses that many dot-coms—from Amazon to Webvan—also required. In short, the assumption was based on a dematerialized conception of ICTs and their roles in business.

These kinds of "disconnected discourses" about ICTs are not just the products of a gullible public. Even relatively sophisticated computer scientists can get swept up in entertaining but distracting debates about such topics as the likelihood that computers will develop superhuman intelligence (including emotional sensitivity) in the next century—and whether we should welcome it or work hard now to avoid that possibility (Joy, 2000; Kurzweil, 1999;

Moravec, 1999). The assumptions made by most of the public participants in these debates extracted an ICT—in this case advanced robots—from the industrial and military social contexts in which they would be most likely to be developed and used. Some of the technologists emphasized implausible domestic settings over the more likely industrial and military settings (Moravec, 1999). Social Informatics research has focused on near term (5–20 year) organizational and social changes, rather than the 100-year perspective of these debates. However, the way that Social Informatics researchers would identify the relevant participants in advanced robotics development and their institutional relationships would dramatically reframe the debates to be more relevant to our likely social futures.

Beyond the popular discourses, there is reliable, evidence-based knowledge about the roles and effects of ICTs in both organizations and, more broadly, in social life. This body of knowledge comes from more than thirty years of systematic, empirically anchored investigation, extensive analysis, and careful theorizing. However, this research is difficult for many researchers and professionals to access. It spans many topics, is published in the journals and books of several disciplines, and draws on a variety of theories and research methods. This collected knowledge provides a rigorous, but also a rich and vivid, basis for understanding the multiple roles that ICTs play in our lives.

Social Informatics is a new name for this body of knowledge. A serviceable working definition of Social Informatics is the systematic study of the social aspects of computerization (a more formal definition is found in Section 1.2 of this chapter). In the rest of this chapter we outline and provide examples of the insights, literatures, and value of a perspective that is grounded in Social Informatics research. However, this is not a textbook or an anthology of Social Informatics: It is a pointer to the practical value of the scholarship on organizational and societal effects of computerization. It is also an argument for and demonstration of the practical value of this scholarship.

The primary goal of this book is to introduce you to Social Informatics research. In doing this, we explain why this body of knowledge is important for all who participate in the design and use of, and education and policy decisions about, ICTs in organizations and society. The book is organized for several audiences involved with ICTs, including:

- Scholars and administrators who are developing and/or reviewing curricular proposals for courses that examine ICTs and social change

- Educators whose teaching and/or research relates to ICTs and social change

- Scholarly analysts involved in debates regarding ICT policies (in campus, local, national, and international venues)

- Program funders who support research about ICTs and social behavior

This list excludes one important audience for the results of Social Informatics research: students of ICT and working ICT professionals. Social Informatics theories are not only aimed at large scale organizational or social change. The research findings, conceptualizations, and theories are also relevant for people who strategize about, develop, or support ICTs in relatively small-scale settings. However, this is not a textbook that would be most readily accessible to these important audiences. They are audiences for the research as it is translated into other books and articles that are specifically written for students or working professionals (e.g., Kling, 2002).

In our view, too few students graduate from the variety of IT degree programs—Computer Science, Information Systems, and Information Science—with a solid analytical understanding of the complexities of IT implementation and use in organizational settings. Some leading consultants willingly admit that 75 percent (a favorite number) of IT-enabled projects fail or dramatically underperform, as most students are not provided with a strong analytical understanding of socio-technical failure modes. Rather, the majority of IT students are taught a technology-centric approach to organizational change, and are too often blind-sided when their well-intentioned project to support a "next big thing" with its own acronym and jargon—TQM, BPR, ERP, KM, Web-communities, or CRM—shows very much less organizational value than proponents had seriously anticipated.

The unambiguous conclusion of the highest quality empirical Social Informatics research is that technology-centered organizational interventions often fail. When they fail, it is rarely a technological failure (though that can happen, such as when prototype systems don't scale up well). Some failures may be in project management. More often, failures seem to be socio-technical—workplace requirements are poorly understood by information systems designers, information systems are not well integrated into pre-existing work flows, information systems are underused because they don't resolve the issues of professionals who are supposed to use them (perhaps they were "best practice" for a different kind of organization) or system use conflicts with organizational incentive systems (a major issue with knowledge management, but even with older concepts, such as expert configuration systems for large computer systems sales support). Well-integrated socio-technical interventions seem to be most workable, though even they are not foolproof.

1.2 Defining Social Informatics

Since the deployment of the first commercial digital computers in the 1950s, their potential power to extend human and organizational capabilities has excited the imaginations of many people. Their potential has also evoked fears that use would lead to massive social problems, such as widespread unemployment. In the 1950s and 1960s, digital computers were relatively expensive (often costing hundreds of thousands of dollars) and relatively few were in use. Consequently, it was difficult to observe their effects, and the writing about computerization was primarily speculative. For example, the concerns about computerized systems becoming efficient substitutes for human labor led to speculation about mass unemployment, radically reduced work weeks, and the "problem" of how millions of people would be able to manage huge amounts of leisure time. From today's perspective, in which computer systems have become ubiquitous and professional work-weeks seem to have expanded, these speculations may seem quaint.

In the late 1960s and early 1970s, some social scientists began empirical observational studies of the consequences of computerization inside organizations. During the 1970s and 1980s, this body of research expanded to cover topics such as the relationship between computerization and changes in the ways in which work was organized, organizations were structured, distributions of power were altered, and so on. Most of the empirical social research was conducted within organizations because they were where the computers and the people who used them most intensively were located. We will discuss the findings of some of these studies in other chapters of this book. Even though these studies may seem to be dated and of limited relevance in the era of the Internet, they can help us understand some key aspects of contemporary issues. Here, it is sufficient to say that some important studies contradicted the prevailing expectations about the effects of computerization that were seen in the books and articles written for ICT specialists, managers, and the broader public.

By the 1980s, research about the social aspects of ICTs was conducted by academics in a number of different fields, including information systems, information science, computer science, sociology, political science, education, and communications. These researchers used a number of different labels for their specialty area, including "social analysis of computing," "social impacts of computing," "information systems research," and "behavioral information systems research." For over thirty years, these research studies were published in the journals of the diverse disciplines, and were written in the researchers' distinctive disciplinary languages. As a consequence, it was hard for many researchers, let alone nonspecialists, ICT professionals, and ICT policy-analysts, to easily track relevant research.

In 1996, some participants in this research community agreed that the scattering of related research in a wide array of journals and the use of different nomenclatures was impeding both the research and the abilities of "research consumers" to find important work. They decided that a common name for the field would be helpful. After significant deliberation, they selected "Social Informatics." (In Europe, the name informatics is widely used to refer to the disciplines that study ICTs, especially those of computer science, information systems, and information science.) Some members of this group held a workshop at Indiana University in 1997, and agreed upon a working definition: *Social Informatics refers to the interdisciplinary study of the design, uses, and consequences of ICTs that takes into account their interaction with institutional and cultural contexts.* Social Informatics analyses that are bounded within organizations, in which the primary participants are located within a few specific organizations, are referred to as organizational informatics. Many studies of the roles of computerization in shaping work and organizational structures fit within organizational informatics.

This definition of Social Informatics helps to emphasize a central principle: ICTs do not exist in social or technological isolation. Their "cultural and institutional contexts" influence the ways in which they are developed, the kinds of workable configurations that are proposed, how they are implemented and used, and the range of consequences that occur for organizations and other social groupings.

Social Informatics is characterized by the problems being examined rather than by the theories or methods used in a research study. In this way, Social Informatics is similar to other fields that are defined by a problem area such as human–computer interaction, software engineering, urban studies, and gerontology. Social Informatics differs from fields such as operations research, where methodologies define their foci and boundaries. Social Informatics research is empirically focused and helps interpret the vexing issues people face when they work and live with systems in which advanced ICTs are important and increasingly pervasive components.

Social Informatics research comprises *normative, analytical,* and *critical orientations,* although these approaches may be combined in any specific study. The *normative orientation* refers to research whose aim is to recommend alternatives for professionals who design, implement, use, or make policy about ICTs. Normative research has an explicit goal of influencing practice by providing empirical evidence illustrating the varied outcomes that occur as people work with ICTs in a wide range of organizational and social contexts. For example, some early research (e.g., Lucas, 1975) showed that information systems were much more effectively utilized when the people who worked with them routinely had some voice in their design. One approach, called participatory design, was built on this insight, and researchers tried to find different ways that users could more effectively influence the designs of systems that they used. Further, some of these studies found that it was important to change work practices and system designs together, rather than to adapt work practices to ICTs that were imposed in workplaces. The recommendations from this body of research are rather direct: ICT specialists and managers should not impose ICTs on workers without involving them in shaping the new ICTs and the redesign of their work practices. These recommendations differ substantially from the strategies of some business reforms of the early 1990s, such as Business Process Reengineering (BPR), whose advocates preferred that ICTs and work be designed by people who were not invested in the workplaces that were being changed. Social Informatics researchers blame some of the failures of BPR on an ideology that undervalues workers' knowledge about their work.

The *analytical orientation* refers to studies that develop theories about ICTs in institutional and cultural contexts, or to empirical studies that are organized to contribute to such theorizing. Analytical research develops concepts and theories to help generalize from an understanding of ICT use in a few particular settings to other ICTs and their uses in other settings. For example, one line of analysis examines specific ICTs as embedded in a larger web of social and technical relationships that extend outside the immediate workplace (or social setting) where the ICTs are used (Kling, 1993; Kling & Scacchi, 1982). This line of analysis indicates that complex ICTs may be workable where technical support is available "in the environment." Thus, public schools in university towns may be able to use more complex ICTs when technically skilled undergraduates provide technical support through part-time jobs or independent study courses. The same ICTs may prove unworkable for public schools in cities where inexpensive technical talent is unavailable. The analytical approach, in this case, examines the way that the social milieu is organized to provide resources for training, consulting, and maintaining ICTs, rather than simply the technical simplicity/complexity of the ICT in social isolation.

The *critical orientation* refers to examining ICTs from perspectives that do not automatically and uncritically accept the goals and beliefs of the groups that commission, design, or implement specific ICTs. The critical orientation is possibly the most novel (Agre & Schuler, 1997; Schultze & Leidner, 2002). It encourages information professionals and researchers to examine ICTs from multiple perspectives (such as those of the various people who use them in different contexts, as well as those of the people who pay for, design, implement, or maintain them) and to examine possible failure modes and service losses, as well as ideal or routine ICT operations.

The critical orientation is exemplified by the case of some lawyers who wanted to develop expert systems that would completely automate the task of coding documents used as evidence in civil litigation. Social Informatician Lucy Suchman (1996) examined the work of clerks who carried out this coding work and learned that it often required much more complex judgments than could be made by rule-based expert systems. She recommended that information systems be designed to help the clerks with their work rather than to replace them.

Ina Wagner (1993) examined the design of a surgical room scheduling system and found that major stakeholders (surgeons, nurses, and patients) had somewhat conflicting preferences. If a system were to be designed, the designer would have to take sides in a set of workplace disputes. Ann Rudinow Saetnan (1991) found that an automated surgical room scheduling system was being used only as a record keeping system because of conflicts between surgeons and nurses about when to make exceptions to the automated schedule. These studies indicate that a systems designer who tries to develop "a better automated scheduling system" may have trouble in designing for only one group, such as surgeons.

An important set of instances arises in the analysis of the safety and effectiveness of systems for people and the operations of organizations. For analysts who conduct post-mortems on ICTs that have failed, it is common to find that the designs or implementations of these systems were not critically examined for the variety of conditions under which people might use them or the ways that they could interact with other limitations in the technical or social systems in which they were embedded (Kling, 1996c; Neumann, 1995; see also Chapter 2, Section 2.1 for further discussion and examples). The findings of Social Informatics research would lead an informed analyst to frame the discussion of a new or changing ICT within a detailed depiction of the organizationally-situated social conditions of likely use.

1.3 The Value of Social Informatics

The empirical base of Social Informatics research provides valuable insights into the contemporary issues with computerization. Some examples that we will discuss later in this book include:

- How can we best understand the meaning of "access to the Internet" in ways that help to foster policies to reduce the "Digital Divide"?

- When does the reliance on weapons systems that use advanced ICTs risk escalating a war rather than reducing conflict?

- How can organizations effectively use computer networks to help their professionals share important information about expensive projects?

- To what extent and when have ICT developments fostered "paperless offices"?

- When can ICTs in K–12 classrooms replace traditional media, such as textbooks, and when are such substitutions likely to be costly and pedagogically troublesome?

One reason why many predictions about the social effects of specific ICT consequences have been so inaccurate is that they are based on oversimplified conceptual models of specific kinds of ICTs or of the nature of the relationship between technology and social change. For example, a simple and common way to view the role of ICTs is as a set of discrete tools. In this view, the computer is a machine that can help rapidly produce a thick book in a few minutes or rapidly solve a complex differential equation. ICT applications like these, wondrous as they are, take on an added transformative dimension when they are networked with other information technologies, such as those that enable people to use the World Wide Web to get up-to-date weather reports or make it easier for a team of scientists to work together even when they are located in different time zones. Further, assumptions about these relationships and models are often tacit, making them even more powerful because they are taken for granted. For example, many armchair analyses of computerization assume that:

- ICTs have direct effects upon organizations and social life.

- These effects depend primarily upon the ICTs' information processing features.

- The information processing features of new ICTs are so powerful relative to preexisting technologies that they effectively determine how people will use them and with what consequences.

For example, the U.S. national effort to "wire" K–12 public schools to connect to the Internet reflects a belief that students' access to the Internet will improve their educations. The motivation behind this reasoning is laudable. An analysis that pushes beyond the face value of this belief leads to questions about how this wiring will actually be done and what changes in the educational process will lead to improved learning. For example, many primary and secondary teachers do not know how to use Internet resources to extend their class activities (and will require both training to get prepared and ongoing support to maintain competence). Further, most schools' computers are in special labs, so that the computing is not integrated into routine classroom practices. Instead, and by design, the computing is often isolated from the curriculum. Thus, the potential value arising from the technical triumph of wiring the school is overshadowed by the need for changes in teacher training and support and to the large-scale curricular (and floor plan) design in order to incorporate computing. And, even after these changes, the issue of exactly how Internet use improves learning has not been addressed. (We examine this topic in more detail in Chapter 3.)

The body of empirical research in Social Informatics does not make these tacit assumptions about the roles and uses of ICTs. In fact, this research has shown that many forms of ICTs, such as groupware, instructional computing, and manufacturing control systems, are often abandoned or reshaped to be used in new ways. In addition, many ICTs create problems that their designers and advocates did not effectively anticipate.

Further, the Social Informatics research literature shows that the consequences of ICT use can appear "contradictory" because they can differ across the various situations in which the ICTs are deployed. Some "distance education" courses taught over the Internet are found to be distressing to their student participants, whereas others develop more positive learning environments (Hara & Kling, 2002). Sometimes computerization leads to organizational decentralization and at other times to centralization of control. Sometimes computerization enhances the quality of jobs and other times jobs are degraded through tightened controls and work speedup.

This book identifies some of the ideas that have come from over thirty years of Social Informatics research—systematic and empirically grounded research about the design, development, uses, and effects of ICTs in social life. Because these findings draw from multiple disciplines and are couched in the specific and particular scientific languages of these disciplines, relatively few of these ideas have been disseminated effectively and, consequently, have not shaped the working practice of most information professionals. Further, much of the body of Social Informatics knowledge has not yet been integrated into many curricula to help better educate young ICT-oriented professionals, and has yet to influence research in related areas, such as digital libraries and new forms of organizing.

As we introduce you to Social Informatics research, we hope to provide you with a useful point of entry into this research world. In the chapters that follow, we discuss the meaning of the concept of Social Informatics and the theories, approaches, and findings that characterize Social Informatics research. We also explain how Social Informatics can be integrated into the curricula of programs and courses focusing on ICTs and social and organizational change.

Endnote

1. The acronym "ICT" refers to information and communication technology—artifacts and practices for recording, organizing, storing, manipulating, and communicating information. Today, many people's attention is focused on new ICTs, such as those developed with computer and telecommunication equipment. But ICTs include a wider array of artifacts, such as telephones, faxes, photocopiers, movies, books, and journal articles. They also include practices such as software testing methods, and approaches to cataloging and indexing documents in a library.

CHAPTER 2

The Consequences of ICTs for Organizations and Social Life

Scholars, information professionals, policy analysts, and the public are each, in their own ways, working to understand the consequences of ICTs on their lives, institutions, and social life. For example, the educational and social consequences of school leaders enabling their students to use the Internet is a topic of discussion among politicians, educators, and parents, as well as researchers. In contrast, topics such as whether enterprise integration systems will enable organizations to streamline their operations or will become costly forms of "electronic bureaucracies," are of more specialized interest, mostly to information professionals and managers in large firms and to researchers. Whether the forms of computerization are those that most directly concern specialized professionals or are of broader public interest, the consequences of computerization are a pervasive concern for society.

Understanding the organizational and social consequences of having, using, and developing ICTs is increasingly important for contemporary professional practices and social policies. As we have noted in Chapter 1, researchers, professionals, consultants, journalists, and pundits have produced a large and growing body of writing about these topics. However, this writing is often difficult to access, much less to comprehend. Moreover, we consider the most reliable portion of these writings to be that which is produced by researchers who study these topics.

This body of literature is known as Social Informatics and in this chapter we begin outlining some of the common findings. To help frame this outline, we begin by discussing why direct effects theories of the consequences of computerization have not been very effective. Direct effects theories predict that computerization leads to some directly attributable outcome. One pervasive form of the direct effects model is technological determinism. Technological determinism treats ICTs as information processing systems whose technical characteristics *cause* specific social changes when they are adopted and used.

Although technological determinism can be applicable and useful in situations that are characterized by high degrees of control and short time frames, it has limited value in dynamic and complex situations that unfold over longer periods of time. Technological determinism cannot adequately account for the interactions among ICTs; the people who design, implement, and use them; and the social and organizational contexts in which the technologies and people are embedded.

There are alternatives to technological determinism and other direct effects theories. Typically these alternatives to direct effects approaches are socio-technical in nature (MacKenzie & Wajcman, 1999) and are often more complex. By socio-technical we mean that they attempt to conceptualize both social and technical aspects as interrelated (Mumford, 2000).

Socio-technical theories help to make clear that the relationships among the features of a particular technology (in our case ICTs) are intimately tied to the specific individual and institutional arrangements among people and their larger social milieu. For example, one of the simplest and most widely used forms of socio-technical analysis is systems rationalism (Kling, 1980). Systems rationalism is a perspective that conceptualizes ICTs as rule-bound and carefully structured and then generalizes these rule-based characteristics to people, groups, and organizations. From this perspective, organizations and the people who work in them constitute rational systems, with formal common goals and work practices carefully designed to meet these goals. These systems can be analyzed at varying levels of granularity in terms of the costs and benefits of alternative sets of goals and practices.

Systems rationalism is a useful starting point to help understand the value of ICTs in organizational practices, social activities, and work life, but it is not a good endpoint for such analyses. Like all analytical models, systems rationalism simplifies the nature of the ICTs, the nature of people and their relationships, and the ways in which people interact with technologies. Beyond simplification, systems rationalism is problematic in that it tends to emphasize formalities. For example, people's work is represented by their formally defined tasks, such as a journalist's work tasks being formally defined as conducting interviews and writing articles. This differs substantially from a worker's *actual* tasks, which, in the journalist's case, may include such things as borrowing from other news stories, spending much time figuring out whom to interview, updating PC software, or scavenging for a new printer cartridge in order to print out a story draft. The vast discrepancy between a formal listing of job tasks and how employees actually spend their time also includes the issue of how formal job descriptions overlook the amount of dependency that workers have on other people.

By focusing on formalities and on simplifications, systems rationalism theorists often depict as streamlined processes that are much more messy and complex. This means that systems rationalism may be useful primarily in cases where the ICTs are designed and implemented to resolve a narrow set of well-understood organizational problems and there are high levels of consensus about problems and solutions by the major participants. However, this is not all that different from direct effect theories.

One of the critical steps to a better understanding of the roles that ICTs play in organizational and social life is that use is tied in with the messy and complex way in which people work and live. For example, Suchman (1987, 2002) shows that using ICTs demands extensive micro-coordination among people and machines and that this is often difficult to foretell. Woods and Patterson (2001) find in their study of high performance systems (like mission control and nuclear power plant operations) that the work is dependent on ICTs and that these systems often cannot meet the needs of the workers (or the task) when most needed. That is, under increased stress, the ICT demand more complex work-arounds, not less. Simply, socio-technical models of ICT use lead to a broader understanding of how computerization is engaged and what its effects are.

Although socio-technical models of ICT use often provide deeper analytic insight, they take longer to conduct and explain than do direct effects approaches to understanding ICT use. The additional time is driven in part by the need to spend time to understand the work practices, uses of ICTs, and the organizational and social structures in which these work practices and uses are embedded. The additional time and effort and the rigorous methods of collecting and analyzing the range of data gathered typically lead to this work entering the scholarly and public discourses after the punditry and armchair theorizing. Thus, much of the Social Informatics research appears to contradict common perceptions. A more careful depiction would be that these common perceptions of ICT use are often built on crude and misguided assumptions about both the ways in which people work and the outcomes of using ICTs.

This short review of direct effects and socio-technical models of ICTs helps to make clear a conflict between the practical importance of being able to make predictions about the consequences of using ICTs and the knowledge that predictions may be fairly unreliable. Much of the reliability and robustness of consequential predictions depends upon the following:

- The time frame (shorter term is often more reliable than longer term prediction)

- The units of analysis (it is easier to predict average behavior for many groups or people than it is to predict the behavior of specific individuals)

- The level of prediction (it is easier to make "rough cut" predictions about generalized organizational or social consequences of technological change than more fine-grained predictions)

An example, particular to the third point, is that it may be easier to assess whether a new electronic forum will improve the quality of discussion in a professional association than to predict the number of association members who will participate routinely.

In the rest of this chapter, we present general findings drawn from Social Informatics literature (see Table 2.1). In the first section, we focus on the social aspects of the socio-technical nature of ICTs. In the second section, we focus on the technical aspects of the socio-technical relationship. In the third section of this chapter, we provide an institutional frame that provides context to support our socio-technical analyses.

Table 2.1 Social, Technical, and Institutional Nature of ICTs

Social Nature of ICTs
　　ICTs are interpreted and used in different ways by different people
　　ICTs enable and constrain social actions and social relationships
　　ICTs provide a means to alter existing control structures
　　ICTs can lead to negative consequences for some stakeholders

Technical Nature of ICTs
　　ICTs have both communicative and computational roles
　　ICTs have temporal and spatial consequences
　　ICTs rarely cause social transformations
　　ICTs are not magic bullets: they do not solve things by themselves

The Institutional Nature of ICTs
　　ICTs social and technical consequences are embedded in institutional contexts
　　ICTs often have important political consequences

2.1 The Social Nature of ICTs

In this section, we discuss three general findings from Social Informatics literature that highlight the social (and organizational) nature of ICTs.

2.1.1 ICTs Are Interpreted and Used in Different Ways by Different People

One of the simplest conceptions of an ICT (or a service provided via the use of an ICT), such as e-mail, a specific digital library, a project scheduling system, etc., is that it embodies the same meanings for all its users. However, Social Informatics researchers have found that people frequently interpret and interact with ICTs in more complex and varied ways (see Kling, 1980, 1993; Newell, Scarbrough, Swan, & Hislop, 1998; Orlikowski, 1993).

A case study of Lotus Notes' use at a major consulting firm illustrates this idea.[1] This consulting firm, which we will call Alpha, bought specialized equipment and 10,000 copies of Lotus Notes for its staff in 1989.[2] Lotus Notes, a documentary support system, is superficially similar to an Internet-like system with bulletin boards, posting mechanisms, discussion groups, and electronic mail for organizations. Depending upon how Notes is used, it can act as an e-mail system, a discussion system, an electronic publishing system, and/or a set of digital libraries.

Alpha is an international consulting firm with tens of thousands of employees worldwide. Its director of Information and Technology believed that Lotus Notes was such a powerful technology that its usefulness would be self-evident, and that the main thing to do was to rapidly roll it out to the consulting staff and let them use it in order to find creative ways to share information. The director felt that Notes was such a valuable tool that people would not need to be shown illustrative business examples for its use, and that providing examples would be counterproductive as it might stunt employees' imaginations. He felt that the consultants should simply be given an opportunity to use it, and they would learn how to use it in creative ways.

The director of Information and Technology was concerned that his firm was employing thousands of "line consultants" in different offices all over North America who were working on similar problems, but rarely sharing their expertise. Sometimes a consulting team in Boston would be dealing with an issue very similar to one being handled by a consulting team in Toronto or San Francisco. The consultants had no easy way of sharing their solutions to the problems they were facing with consultants in other offices. The plan was that the firm's line consultants would use Notes to store what they knew, and then share it.

The first test of Notes was with the information technology staff. They tended to use Notes; they found it interesting and used it fairly aggressively for sharing information about their own projects. Alpha's tax consultants in Washington, D.C. were the second group that used Lotus Notes. The tax consultants disseminated tax advisories to Alpha offices around the country about shifting changes in tax legislation that might affect their clients. Alpha's tax consultants made substantial use of Lotus Notes to broadcast their tax advisories.

The line consultants were intended to be Lotus Notes' primary users. However, organizational informatics researchers found that the senior line consultants, who were partners in the firm, tended to be modest users, while the more numerous junior line consultants, called associates, were actually minimal users. They often seemed uninterested in learning how to use Notes, readily gave up if they faced early frustrations with Notes, and as a group did not spend much time with it. Here we have a pattern of different groups within an organization having different practices in working with Notes. How can we explain such differences?

One explanation focuses on the incentive systems in the firm. A good place to start our analysis is with the associate consultants and the partners. Alpha—and many other large consulting firms in North America—reviews its consultants through a demanding promotion system. The associates receive an "up or out" performance review every two years. In the first few rounds at major firms, about half of the associates are fired at each review. In this "up or out" system, the goal of many of the associate consultants is to be promoted to the status of partner.

The associates are valued for their billable hours, and are effectively required to bill almost all of their time. As they become more senior, their ability to attract new business becomes more critical. "Billable hours" means that they have an account to which they can charge their time. Lotus Notes, the revolutionary technology, was not provided to them with a "training account" to which they could bill their time. Consultants who wanted to use Notes had to have an account to charge their time against, and the initial learning time was in the order of 20 to 30 hours. In 1991, the consultants were billing at about $150 an hour, so they had to find a client who would be willing to pay $3,000 to $4,500 for them to learn a system whose value wasn't yet clear to them. Many had trouble justifying that amount of expenditure to any of their clients at the time that they were participating in the Notes rollout. There was also an important question as to what the consultants would actually do with Notes after they learned how to use it. Consequently, relatively few associates saw much value in Notes, and there were no exemplary demonstrations showing them how other successful line consultants used Notes.

On the other hand, partners have substantial job security (similar to university tenure). They could afford to experiment with Notes. They were more willing to invest some time to explore, often using e-mail, occasionally developing and sending memos, and so on. Overall, this case study contradicts the popular "Nintendo generation" explanation: "In the future, we don't have to train people about computing, because the Nintendo kids (or the Net kids) will learn quickly." In this case the younger consultants generally had less incentive to learn Notes than did the middle-aged and older partners.

What about the information technology staff and the tax consultants? These groups also had an advantage in their forms of job security. Many of the information technology staff were technophiles who liked to work with interesting new applications. Lotus Notes has been helpful for people who can invest time in learning how to use it, especially when they have joint projects and substantial motivations for communicating, for documenting work, for sharing memos, and so on.

The tax consultants, who were located in Washington, D.C., had significant incentives to show their visibility in and value to the firm. Their fixed salaries were not based on billable hours, and this allowed them more freedom to explore Notes' uses. Lotus Notes allowed them to broadcast for visibility. It gave them the ability, in effect, to electronically publish their advice and make it quickly available to many of the consultants around the firm who wanted to read the Notes database. They hoped it would enhance their visibility, and thus show that the Washington office was not just overhead, but an important, contributing part of the firm.

In short, although they proved to be of considerable importance, organizational incentive systems were not part of the original marketing story of Lotus Notes. It was the interesting information processing features enabled by Lotus Notes that were emphasized in numerous stories in the technical press (see for example, Kirkpatrick, 1993).

This case illustrates varied consequences of Notes' use in a large consulting firm, as opposed to one fixed effect. Finding such varied, conflicting consequences in different settings is common in this body of research. Social Informatics researchers do not simply document the various consequences of computerization, but also to develop empirically grounded concepts that help us to predict (or at least understand) variations in the ways that people and groups use information technologies (Lamb, 1997; Robey, 1997). For example, analysis of the different organizational incentive systems for different professionals increases understanding of the disparate outcomes in the case discussed here. (Also see Markus and Keil, 1994, for a case study of a little-used, large-scale expert system whose use was not supported by organizational incentive systems.)

One key idea of Social Informatics research is that the "social context" of ICT development and use plays a significant role in influencing the ways that people use information and technologies, and thus influences their consequences for work, organizations, and other social relationships. Social context does not refer to some abstracted "cloud" that hovers above people and ICTs; it refers to a specific matrix of social relationships. The social context may also be characterized by particular institutional relationships. For example, one set of institutional relationships would be the incentive systems for organizing and sharing information at work. In the case just described, different groups within Alpha have different incentives to share information about the project "know-how," and this affects their use of Lotus Notes.

The use of the World Wide Web (or Web or WWW) in university instruction today is another contemporary example. A growing fraction of university faculty are eager to explore the Web as a way to enhance some aspect of their teaching—whether it is making class materials more readily available to their students, developing online discussions, or devising new forms of interactive activities. Many other faculty, especially nontechnical faculty at research universities, are much less interested in working with the Web because they believe that using it would require a lot of their time to learn HTML and other ICTs, to develop and maintain materials, and so on. These stances are further mediated by the extent to which the faculty's departments provide technological help or teaching assistants to help with the development of Web materials. There are many other contingencies that can influence the ways that university faculty interpret the Web in relation to their teaching. The main point is that these interpretations are not uniform—people's interpretations of an ICT are based on prior beliefs, and the perceived new opportunities and demands it creates relative to their other opportunities and commitments. How people interpret an ICT is of considerable importance because people (and organizations) with different interpretations will adopt and use ICTs differently.

2.1.2 ICTs Enable and Constrain Social Actions and Social Relationships

ICTs are sometimes called "technologies of freedom" because they extend the abilities of people and organizations to access data, communicate, and so on. It is common for many technology-centered accounts of new ICTs to emphasize the ways that they enable new kinds of action that were previously more costly, difficult, or impossible. However, many of the ICTs' freedoms come with some less visible constraints.

The shift from paper to electronic documents offers some interesting examples. People who work with paper documents often face the constraints of needing to travel to a library to obtain them. However, once they are in hand, they can be read virtually anywhere. In contrast, digital libraries open the possibility of having documents accessible on one's desktop. However, unless they are then printed out, people can't easily read them on planes, in bed, at the beach, and so forth.

Researchers frequently find that ICTs, in use, do not simply "open new possibilities" for organizational action, for organizing work, for professional communication, for supporting education, and so on. Rather, they restructure existing information processing and the social relationships which surround these (Sawyer, Crowston, Wigand, & Allbritton, 2003). For example, a work team may use group calendaring software to help coordinate its activities.

One advantage is that the team's members may be able to more easily schedule meetings. However, team members must also bear additional responsibilities to keep their own schedules up-to-date, and thus to log in as they schedule and reschedule meetings and other activities during their workdays (Pino & Mora, 1998).

The ways in which developers and local implementers configure ICTs to restructure social arrangements can take place at the level of whole organizations as well as in smaller scale settings such as work groups. To manage increasingly complex supply chains, streamline business processes, and coordinate resources around the world, a growing number of the world's biggest corporations are implementing enterprise-wide IT management systems that are supposed to facilitate sharing data and tracking operations both around the world and across functional areas. As we noted previously, enterprise systems are touted as a means of reducing the operational and maintenance costs of running stand-alone systems and providing an opportunity to implement real global change. Today, many large business firms are investing hundreds of millions of dollars each in enterprise systems. These systems may be spoken of as tools to help people manage more effectively. But, in practice, each enterprise system tends to impose a more centralized information structuring regime on a firm.

Some managers have found enterprise systems to be congruent with their current management practices, whereas others have restructured their businesses to fit the constraints of their new enterprise systems. A few organizations have abandoned enterprise integration projects after spending up to $500 million because the gains from the systems did not seem worth the new constraints (Davenport, 1998). Some of these firms reverted to information systems whose architectures better supported their more decentralized ways of managing and working. In a few cases, the firms embarked on developing new enterprise system projects whose technological architectures seemed to offer a more appropriate set of opportunities and constraints.

2.1.3 ICTs Provide a Means to Alter Existing Control Structures

When first introduced into social and organizational settings, the relationship between ICT use and social structures is reciprocal. Organizational informatics researchers have found that ICTs can restructure workplaces through the ways in which they are incorporated into the everyday lives of those who use them (Barley, 1986, p. 81). Technologies are also shaped by the everyday actions of those who routinely use them and the social settings within which they have been implemented (Orlikowski & Robey, 1991, p. 151). However, organizations come to stabilize around some configurations of work practices and ICT configurations. Thereafter, changes are incremental until there is some substantial "outside change"—such as changing physical locations, a shift to a new kind of ICT, a major shift in the mix of work or services being produced, and so on.

Once organizations stabilize around some technological configurations (especially standards for complex infrastructures such as networking protocols, operating system families, and databases), they become taken for granted and institutionalized in ways that impede other subsequent innovations (Kling & Iacono, 1988). For example, in the early 1980s, mainframes or minicomputers running large database management systems and analytical programs and connected to workers via "dumb terminals" dominated the organizational computing landscape. Few central information systems organizations supported PCs or encouraged their use. Many departments obtained their first PCs when professionals and managers "snuck them in"— by hiding their purchases under safe accounting categories, such as "word processors" or "engineering instruments," or even by physically carrying their own PCs into their offices.

The range of contemporary policies about access and use of the Internet and e-mail reflects the ongoing efforts to create institutional stability. That is, computerization efforts (whether they achieve their intended goals or not) affect the ways in which people organize and interact. Thus, current controversies about Internet use reflect this same tension between institutional stability and technological innovation. For instance, librarians must balance community opinions of decency against patron's first amendment rights to access indecent materials at public computers in local libraries. Policies regarding the use of organizational computers to access certain Web sites, send e-mail of certain types, and do other things online vary by both organization and status within organization. Further, the rapid growth of peer-to-peer file sharing and the problems with enforcing intellectual property rights in the face of widespread digital music sharing is destabilizing the music industry (and threatening many other industries whose revenue is based on trade of intellectual property).

Control structures are often built into ICTs and these often enforce gender, race, and socioeconomic class distinctions. For example, as Hoffman (1999) notes, in the early designs of word processors, designers provided fixed-path menu structures and rigid document controls. Designers believed that the users (primarily women) would need this extra control because they did not understand computers. However, the users were experts at typing, and the enforced control structures inhibited their ability and reduced their productivity. Hoffman further notes that, as word processors were redeveloped for knowledge workers (primarily using masculine views of knowledge workers) the control structures were minimized.

2.1.4 There Can Be Negative Consequences of ICT Developments for Some Stakeholders

New ICT developments are usually promoted by their sponsors in terms of their foreseeable, direct benefits to some groups. However, it is common when mobilizing support for them to downplay or ignore their disadvantages. For example:

- Many large business firms introduced Enterprise Resource Planning (ERP) systems that could help them to streamline their information flows, solve Year 2000 problems, save support costs by having a common systems infrastructure, and provide a '"platform"' for future computing growth. However, deploying these systems has led to significant centralization in some firms, as well as disruption (Davenport, 1998; Markus, Axline, Petrie, & Tanis, 2000).

- Many business firms, large and small, are encouraged to develop Web sites as a way of enhancing marketing and increasing their sales. However, many are losing significant amounts of money on their efforts at electronic business (Grover, 1999; Nelson, 1999)—with resulting stress on their staff.

Many new developments have some negative consequences that key stakeholders did not want. Lynne Markus's (1994) careful case study of the social consequences of e-mail use between the staff at the headquarters of an insurance firm (HCP) helps to illustrate this idea. She found that HCP's upper managers required their staffs to rely upon e-mail, and it was the major medium for internal corporate communications at the time of her study. HCP's staff used e-mail to speed communications and bring people closer together. But they also reported significant negative effects, such as feeling that HCP was a less personal place to work. Electronic mail was often used to avoid face-to-face confrontation or unpleasant situations, and often in-office visitors were ignored while employees tended to the demands of messages on their computer terminals.

Systematic use of ICTs changes the nature and extent of social life in ways that may improve some kinds of social relationships and activities at the expense of others (Wellman, 2001). One common and typically unintended outcome of computerization efforts are systematic political repercussions: the new ICT leads to "winners and losers" (Danziger, Dutton, Kling, & Kraemer, 1982; Markus 1981, 1983). For example, Kraut, Dumais, and Koch (1989) show how repeated attempts to support phone operators with automation led to an unintended reduction in the quality of their work life.

Sometimes negative repercussions from computerization are relatively minor, and the benefits of computerization may be of much larger value to many participants. Sometimes the negative repercussions loom large for people who have the power or influence to block a computerization project, or to alter the ways in which systems are used. As an example of the latter, Saetnan (1991) studied the attempts of designers of a surgical room scheduling system, PREOP, to have it adopted in Norwegian hospitals. PREOP was supposed to make more efficient use of staff time and physical resources by automatically scheduling arrangements for operations. However, Saetnan found that the scheduling algorithms did not effectively mesh with the work of doctors and nurses. After some frustration, the staff used the computer system simply to record schedules rather than to optimize them. The nurses had significant influence in scheduling the rooms, and were reluctant to submit to the surgeons, whose preferences were reflected in the design of the new system. Saetnan notes:

> Having learned to master PREOP by overriding parameters, nurses could force PREOP to set up schedules exactly as they had before. ... PREOP became a slave to the old routines and thus reinforced them (Saetnan, 1991, p. 434).

Although the creation of "losers" from the implementation of ICTs may be done purposefully on some occasions, or considered an unavoidable side effect at other times (such as with the phone operators), there are times, as at HCP, when no one foresees the negative impact that some people will experience from computerization. No one "wanted" HCP to feel like a less personal workplace. Even so, HCP's employees integrated e-mail into their work lives in ways that sometimes added barriers between themselves and their colleagues.

One group may also be able to force another group to use an ICT in a way that disadvantages them (Kling, 1980, 1983; Markus, 1981). For example, Clement and Halonen (1998) examined how user groups and information systems specialists in a large utility company differed in their conceptualizations of a "good" system and good systems development practice. According to Clement and Halonen, for the users, a good system should be customized for different offices such that use required minimum knowledge of office-specific codes. This would require code customization for each office, resulting in the existence of multiple unique versions of the software and a great deal of ad hoc programming. From the information systems perspective, good systems development practice requires standardization, version control, and minimal code changes. This approach would lead to the creation of a less user-friendly system that requires more end-user expertise.

Clement and Halonen's study showcases the ways in which different conceptualizations of a good system and good systems development practices benefit one group more than another. User groups favored the ad hoc, customized development practices because they resulted in a more flexible, customizable, user-friendly system. The information systems group preferred a more structured, systematized approach to development because it resulted in a more easily manageable software product.

As these examples illustrate, the distribution of good and bad effects of ICT use is political; these outcomes are not inherent in the features of the technology. The choice of which groups gain or lose through using an ICT is a political decision and these gains and losses reflect *configurational* actions (those taken to shape the design and use of an ICT) (Fleck, 1994). For example, in one large university, the central ICT staff changed the technical protocols that supported e-mail use in order to make better use of new central mail-server computers. The transition to new mail servers was well advertised. What was not well advertised was that the central ICT support staff was also dropping support from some popular e-mail programs that they had previously distributed. Faculty, staff, and students who used these e-mail programs learned about the loss of support only when their e-mail did not work properly on the new e-mail servers. Their phone calls to a university ICT help center put them in contact with sympathetic staff who were unable to help because they were not taught how to advise about problems with e-mail software no longer being supported.

In retrospect, it appeared that the central ICT staff continued to support and had tested the e-mail packages that they preferred in their own work lives. Some e-mail packages that were preferred by faculty or students who worked at home as much as on campus were not used in the central ICT organization. Support for these e-mail packages was quietly dropped to help simplify the work of the university's ICT organizations. It was possible to configure these programs for the university's new e-mail servers by finding technical tips that were posted on the Web of central support organizations at other universities! But this technical self-help was very disruptive and unnerving for the faculty and students who spent several hours doing it, while their professional communication via e-mail was unexpectedly interrupted. In this example, the faculty, students, and staff who used the e-mail packages that were personally preferred by the central ICT support staff were advantaged while those who relied upon different (and often functionally better-sorted) e-mail packages had their professional lives disrupted.

2.2 The Technical Nature of ICTs

In this section, we highlight findings from Social Informatics literature that focus on the technical consequences of ICTs. In doing this, we note that the ways in which we conceptualize an ICT underlies any discussion of its technical attributes (Orlikowski & Iacono, 2001). In Social Informatics literature, ICTs are typically conceptualized as part of a larger socio-technical ensemble, and their particular use or value is portrayed in functional ways (Sawyer & Chen, 2002). This functional view contrasts with computational or feature-based conceptualizations that highlight the algorithmic or designer-centric representations of an ICT.

2.2.1 ICTs Play Both Communicative and Computational Roles

Proponents of both direct effects and systems rationalist theories of computerization tend to emphasize only some of an ICT-based systems' information processing features, such as the size and contents of the corpus of a digital library or the mathematical approach of a forecasting model. However, ICT-based systems usually play communicative roles as well. The earliest computer-based systems tended to emphasize computational capabilities, primarily for diverse scientific applications and business data processing, and this focus often manifests itself in current conceptualizations of the role of computing. However, groupware applications that are designed to enhance teamwork must be viewed in communicative terms. For example, Guinan, Cooprider, and Sawyer (1997) found that software engineering tools, designed to support computation, were more often used to support communication. Similarly, computer networking, heightened by the multimedia presentations in the Web, highlights the communicative roles of today's advanced information technologies.

Human communication plays a role in systems that emphasize intensive computation, as well: The mathematical models that are supposed to enhance human decision making are interpreted within settings where communication between people is preeminent. Professionals and managers discuss (and negotiate) the assumptions behind mathematical models, their structure, the results of modeling runs, and their meaning (Dutton & Kraemer, 1985).

Sometimes computer systems provide important social communication channels and media. Kling and Iacono (1984) report on large manufacturing firms that used certain computerized inventory control and production scheduling systems, called MRP (Material Requirements Planning) systems. These MRP systems are transaction-oriented ICTs whose data refer to materials (e.g., quantities of parts and subassemblies to be ordered or manufactured by certain dates). Direct effects and systems rationalist theories both emphasize the ways in which the use of MRP can help reduce operational costs by significantly reducing inventories. However, Kling and Iacono found that an MRP system can also serve as a social control system because it gives the staff in various departments information about the detailed activities of those in other departments. For instance, purchasing departments are responsible for ordering parts, and different manufacturing departments are responsible for transforming purchased parts or grouping them in subassemblies. Manufacturing staff can attribute chronic shortages in certain kinds of parts, such as electronic circuits or motors, to the specific employees who specialize in buying that family of parts. Similarly, manufacturing staff can use an MRP system to track the performance of other departments in constructing subassemblies or in performing related activities, such as inspection and testing. Acknowledging the communicative aspect of ICTs is important because it increases our understanding of their uses and consequences by including the *social behavior* of people who use them.

2.2.2 There Are Important Temporal and Spatial Dimensions of ICT Consequences

Some of the popular conceptions of the consequences of ICTs for space and time are misleading. Despite the advertising cliché that "computer networks eliminate space and time,"' the reality is that organizations have been able to reduce only some of the temporal and spatial barriers for work, communicating with clients, and so on. Woody Allen once joked that "history is God's way of keeping everything from happening all at the same time." Using an ICT that seems to eliminate the user's dependence on space and time might seem to offer tremendous convenience for individuals and organizations. However, because these users could also be continually accessible to others—regardless of their locations or time zones—they could be easily inundated with communications and demands. This overloading is particularly likely when people work on multiple projects, with several teams, and support many clients.

Clearly, ICTs are enabling people and organizations to reduce some of the communicational restrictions of space and time—in ways that we do not understand very well (Lee & Sawyer, 2002). Time and space are also important considerations in understanding how long the consequences of ICTs take to show up, and where. The consequences of ICT deployments may be relatively immediate (i.e., develop over weeks or months), especially when their use is essential and requires only a few well defined and easily learned skills. Or their consequences may take time to build. For example, a new digital library or electronic journal may take several years to develop a constituency, especially when use is discretionary and the people who may potentially use it have alternative information sources.

2.2.3 ICTs Rarely Cause Social Transformations

Much of the popular literature about ICTs and social change emphasizes "social transformations" and the ways in which ICTs create new social worlds. Empirically oriented Social Informatics researchers who carefully study ICTs and social change find that the pace of change is relatively slow, and that there are usually important continuities in social life in addition to the discontinuities. For example, bold claims have been made about how ICTs have "transformed work." Guevara & Ord (1996) claim that, "The nature of work is currently undergoing a complete transformation. ... Information technology is underpinning this transformation" typifies a commonplace claim in the business and technology press. Some pundits go further in emphasizing the rise of virtual offices, the replacement of jobs with project-level assignments, the demise of large organizations, and so on.

Certainly in the U.S., work life has changed in many ways in the twenty years since PCs became popular. However, some of the shifts have come from a changing mix of work—from manufacturing production to services. Other shifts are the result of increasing numbers of professional and managerial jobs requiring heavy travel as firms scale up geographically. Kling and Zmuidzinas (1994) found that managers had at least 18 different directions in which they could reorganize work.

In practice, computerization projects play only modest roles in restructuring white-collar workplaces. As Brooks (1987) noted in the context of software engineering, there are no (ICT-based) '"silver-bullets." Ten years later, Guinan, Cooprider, and Sawyer (1997) provided empirical evidence that the presence of CASE tools did not radically change software engineering. Similar findings arise from other industries. Salzman (1989) noted that the increased use of Computer-Aided Design (CAD) in engineering did not remove engineers, as it was predicted to do. Fleck (1994) points out that MRP systems did not restructure work as much as they reinforced traditional roles (shop floor worker, manager, technology specialist).

As Markus and Benjamin (1997) note, workplace change is more multifaceted than is suggested by direct effects models that espouse "ADD ICT AND GET CHANGE." As both Burkhardt (1994) and Liker, Roitman, and Roskies (1987) find in separate studies of organizational change following technical change, long-term changes reflect larger organizational, industrial, and economic forces and have little to do with changes in ICT. Beyond organizational work, Sawyer, Crowston, Wigand, and Allbritton (2003) find that extensive computerization of real estate agents in the U.S., and the growth of house-hunting data available to buyers and sellers on the Web, has not radically restructured the work of these agents.

2.3 The Institutional Nature of ICTs

In this section, we develop an institutional perspective on ICTs. As we explain below, institutions are enduring social structures. We note that ICTs are embedded in, help to shape, and are shaped by institutions. In the rest of this section, we review the various forms of institutions, the nature and means of embedding, and the political nature of their actions in shaping these institutional contexts.

2.3.1 Social and Technical Consequences Are Embedded in Institutional Contexts

Institutions, the most enduring and general type of social organization, can range from highly formal and public organizations, such as banks and governments, to more private organizations such as families, and even to loosely structured organizations such as the soccer leagues to which families might belong (Scott, 2001). Institutional memberships may overlap in complex ways: A banker may have a child in the soccer league being coached by one of his clients. Moreover, most people navigate this institutional complexity, moving in and out of a range of institutional settings in their daily lives, with a great deal of dexterity.

The institutional embeddedness of ICTs means that they cannot be conceived of or discussed outside of particular institutional arrangements. As the Lotus Notes case makes clear, differences in professional practices within one formal organization lead to different responses to the same ICT. This makes discussions of best-practices-with-ICT tricky. That is, the practices situated in institutional context "A" are not always transportable to institutional context "B." For example, Markus, Axline, Petrie, and Tanis (2000) note that, although general patterns are clear, specific practices relative to implementing ERP vary among adopters. Important elements of the ERP (such as the centrality of certain modules and functions) may vary for each implementer.

2.3.2 ICTs Often Have Important Political Consequences

In scientific and technical communities, the term "politics" often has an ambiguous status. When analysts use it to refer to governance processes in the larger society, such as the election of public officials, debate on public policies, development of legislation, and so forth, it can have a positive valence. However, scientists and technologists often view the term "organizational politics" with some distaste, as if political processes are necessarily dysfunctional and inappropriate. In fact, in the 1960s and 1970s, some analysts had hoped that computer models could take "the politics out of decision making" inside organizations as well as within public agencies. This hope was not realized (Dutton & Kraemer, 1985).

Researchers who study organizations have found that organizations do not function simply as task or production systems. In practice, a view of organizations as political networks has added an important dimension to the study of organizations generally and to understanding the roles of ICTs in organizational change in particular (Danziger, Dutton, Kling, & Kraemer, 1982; Laudon, 1974; Markus, 1981; Pfeffer, 1981). When viewed as political networks, organizations can be seen to have governance structures, ways of allocating important resources, and ways of legitimizing their actions. Groups within (and outside) an organization are often trying to influence a specific organizations' governance, allocations, or forms of legitimization. Organizations differ in the forms of their political networks: Some are more autocratic whereas others allow (and sometimes even encourage) conflict between competing groups and coalitions. When ICTs are significant organizational innovations they require money and staff to acquire and implement, and it requires some legitimacy to restructure other people's work to align with the new practices that the ICT is supposed to enable.

For example, e-mail systems enhance professional communication only when people actually use them. In the late 1980s and early 1990s, many managers, professionals, and technologists who were enthusiastic about e-mail tried various ways to make e-mail use appropriate and legitimate in their organizations. From a political perspective, e-mail is not simply a "technical solution" that helps solve communication problems (such as reducing the frequency of "telephone tag" between people who are not always in their offices). Researchers have found that e-mail use can give greater visibility (and thus influence) to "peripheral participants," such as people who work at night or in field offices (Sproull & Kiesler, 1991). In this way, e-mail has "political outcomes"—some people gain influence (and resources) while others do not.

In the mid-1960s, political scientist Anthony Downs speculated about the power payoffs of ICTs in organizations and argued that power would shift toward those who collected information. Organizational informatics researchers have found that many ICTs actually can shift the balance of influence and power in organizations by restructuring access to information, technical staff, and the level of legitimacy that informational resources can bring (Danziger, Dutton, Kling, & Kraemer, 1982). Political processes are not static, and political analysts examine the ways that groups jockey for influence as the nature of the opportunities, stakeholders, coalitions, and so forth change over time (Kling & Iacono, 1984).

One important contribution of the political networks view of the role of ICTs in organizational change is that it helps to deepen our understanding of the motivations for different groups supporting and opposing specific forms of ICT developments. Markus (1981) explained the support and opposition for a new centralized accounting system in terms of perceived power shifts. Her explanation contrasts with that of analysts who view people's support or opposition to new ICTs as a psychological disposition of being either pro-innovation or technophobic. In a rich case study, Markus showed how the professionals who supported a new system were those who expected to gain influence from its use, while those who opposed it expected to lose control over their data through its use.

Hodas (1996) developed a power-based analysis of many school teachers' indifference or opposition to ICTs in K–12 schools. He observed that many K–12 teachers place a high value on having control over their students (such as having quiet, orderly classes). Many computerization projects in K–12 schools can upset teachers' control strategies because they require students to move around within classrooms, or between classrooms and laboratories. In addition, for a variety of reasons, children can develop greater expertise in computer use than their teachers (including time on task), and many teachers fear that their authority through expertise will be undermined.

Kenneth Laudon (1974, pp. 164–166) found that the relative power of central administrators was critical in determining whether and how information systems are adopted in county governments. This insight may help explain some of the implementation failures of enterprise systems (mentioned earlier). For example, groups that have sufficient power and perceive threatened reductions in their influence may thwart the efforts of headquarters-based managers and technologists who try to impose a more centralized managerial regime on them. Today, some of the debates about the value of electronic scientific publishing could be usefully examined from the perspective of expected gains and losses of scientific influence. These power-oriented explanations differ from those explanations that simply focus on the advantages an ICT will afford some group.

Endnotes

1. For a more complete discussion of this case, see Kling, 2000.
2. Alpha's unusual mass purchase of Notes for all of their consulting staff was the subject of several reverential stories in the technical and business press.

CHAPTER 3

Social Informatics for Designers, Developers, and Implementers of ICT-Based Systems

The effects of designing, developing, and implementing ICT-based systems on organizations and other social units are still not well understood—even though they are increasingly important to the organizations' operations and outcomes. Consequences of this lack of understanding show up in the high failure rates of many new ICT-based systems; the slow uptake of others; and issues like user resistance, the need for top management support, and the increasing attention to usability (as a means of making the interface more pleasing to its users). There is evidence of extensive and complex interactions among ICT-based systems and their social and organizational environments. What is not as clear is how system designers, developers, and integrators come to learn about these interactions or account for them in design.

In the rest of this chapter, we summarize the research into these effects, which we call the social design of computing, and lay out some principles for improving the social design of ICT-based systems. To do this, we provide some context to explain how the empirical evidence has remained disconnected from the engineering and managerial knowledge base of most ICT professionals. We then explain what we mean by the configurational nature of ICT-based systems and make a distinction between the usability and usefulness of these systems.

3.1 Understanding the Social Design of ICTs

The social design of ICTs is best understood as arising in response to the problems with designer-centric approaches. In this section, we briefly review the designer-focused approach to designing ICT-based systems, and highlight two challenges to understanding social-design approaches.

3.1.1 The Historical Premise of Designer-Focused Systems

By the late 1950s and into the early 1960s, most people who designed and/or used ICTs were scientists, engineers, technicians, and programmers. The ICT-based systems of the time were primarily designed for specific scientific and military applications. Often the users were also the designers. Or, the users and designers were from the same intellectual community. This made it easy to imagine the design as something the designer would use, and minimized the need to understand a user's perspective on design. In essence, this is a form of "'end-user'" computing that sprang up again in the 1980s when the personal computer became a common part of workplaces.

As computing use broadened, early "commercial systems" were developed for very structured activities such as accounting, banking, and insurance record keeping, for which there were often paper-based and electro-mechanical precursors. These processes were purposefully selected because of their structure. Often, these processes had been the object of extensive work changes to systematize (as the early work-design efforts were known, see Nadler, 1963). These efforts progressed to where the people's contributions to the process were seemingly highly routinized and thus seen as easier to automate. Designers worked to ensure that these highly structured processes were then carefully mimicked by the earliest digital systems. However, they often did this without speaking to users (whose work had been documented by others and who were often seen as having little insight on or value to add to the process). Further, the early success of designers in scientific and military applications did not provide any challenge to the designer-centric approach to developing and implementing ICT-based systems.

Because the people who would use the systems did not have much influence in their designs, many had trouble using these computerized information systems because of design flaws. Some systems printed inappropriate checks (e.g., for $.00 or $.01); others printed individual items on whole sheets of paper or wasted clerical time because they did not effectively group items or sort reports. Even for these highly structured systems, arriving at a useful and usable design was more complex than the designers had anticipated. And, it was often easy to point to the users as the source of problems because what they did in practice was often not what the designers had imagined when developing the ICT-based system.

Since then, the development of computer hardware and software has improved, primarily because of the increasing attention to linking what people do to what the ICT-based systems can do. Still, designing any sort of complex computer system for ordinary (nontechnical) people remains difficult. Certainly ICTs can be extremely powerful and complex, doing a vast array of different things with enormous speed—this is their advantage and to a large extent their appeal. For example, a communications system such as electronic mail can be designed not only to let a person send an asynchronous text message to another, but also to send multiple messages, create mailing lists, respond automatically, forward, save, retrieve, edit, cut and paste, add attachments, create vacation messages, fax, and so on. However, if the design is not handled extremely well, people will have to figure out a bewildering array of options in order to use their ICTs. For example, simple mistakes in e-mail (such as an individual using reply-to-all instead of reply-to-sender, or a listserv owner setting reply-to-all instead of reply-to-sender as a default) can have massive consequences (from increased load to embarrassing accidental responses). Further, some people will spend undue amounts of time trying to learn to use some of the system's capabilities at the cost of productivity. Others will simply forgo using features that they can't comprehend, and thus will lose some of the value of their ICT. Overall, evidence shows that relatively few people get the full value that they could and should get from their ICT-based systems (Computer Science and Telecommunications Board, 1998).

3.1.2 The Configurational Nature of ICT-Based Systems

By configurational, we mean that an ICT-based system's uses are not fully inscribed in its design. This stands in contrast to a screwdriver or automated card reader, the uses of which are clearly inscribed in their design. To configure an ICT-based system typically means making compromises between idealized and enacted views of what is being supported. Moreover, configuration is ongoing and continuous, as it is part of designing, implementing, and using ICTs. This ongoing need for compromise reflects the need to balance the way people want to work with the way the system lets them work. The premise is that designers of ICT-based systems are not grappling with well-formed problems but with "messy, indeterminate situations" (Schon, 1987). The extra time and effort needed to adjust various aspects of ICT-based information systems is just one example of the ongoing problems resulting from their increased complexity and potential power. For example, Patterson, Cook, and Render (2002) highlight the difficulties in automating medical patient records. In their studies of U.S. Veteran's Administration Hospital's efforts to digitize patient records, Patterson et al. show how linear views of nurses' work (designed into the record) are at odds with the way nurses actually enact their jobs. Thus, nurses are now spending more of their time working around the system designed to reduce their workload.

A second example of the configurational nature of ICT-based systems builds on the current state of information storage and use in larger organizations. Most of the information in large organizations is spread across dozens or even hundreds of separate computer systems, and stored in various business functions, regions, factories, or offices. Maintaining these many different information systems is perceived as "one of the heaviest drags on business productivity and performance now in existence" (Davenport, 1998, p. 123). For managers who have struggled at great expense and with great frustration with incompatible information systems and inconsistent operating systems, enterprise integration systems, like SAP's R/3 package, promise the "seamless integration" of all the information flowing through a company. In short, these "enterprise systems appear to be a dream come true" (Davenport, 1998, p. 121).

Seamless integration is a powerful, direct effect concept. However, enterprise integration systems are profoundly complex pieces of software. At their core lies a comprehensive database that holds data collected from, and feeds data into, modular applications supporting practically all of a company's formalized business activities. When new information is entered in one application, related information can be automatically updated. Although the marketers of enterprise integration systems promise financial and productivity benefits through flexible configurations, the actual design of these systems often reflects a set of standardized assumptions about the way organizations and businesses should operate. As Davenport (1998, p. 125) notes:

> Vendors try to structure the systems to reflect best practices, but it is the vendor, not the customer, that is defining what '"best"' means. ... As a result, most companies installing enterprise systems will need to adapt or even completely rework their processes to fit the requirements of the system.

Enterprise systems are a powerful example of the configurational nature of ICT-based systems because their design is such that much of the modification occurs during and after implementation (Sawyer, 2001). For example, in order to implement SAP R/3, there are more than 8,000 template, table, and processing decisions that must be made. Often these decisions are made implicitly because few designers and even fewer implementers know what the cumulative effects of all these decisions might be.

Independent of the implementation, design itself is complex and indeterminate. The design process for large-scale systems usually involves a team (or team of teams) of designers from different disciplines (Sawyer, Farber, & Spillers, 1997). The team must interact with the people who will use its product and with the much larger organization of which it is a part. The design and implementation of ICTs typically takes place under considerable limitations of time and resources; it seems unlikely that this will change. Many systems designers are taught to separate organizational issues from technical issues, but in practice, this separation is rarely feasible for designing workable systems. In short, the design of technology-based products is inextricably entwined with social and organizational dynamics.

3.1.3 Usability Is a Partial Response to Designer-Focused Approaches

Designers face numerous and increasingly complex challenges in crafting ICTs that will work well for the organizations and people using them. Organizations are increasingly interested in measurable paybacks from their technology investments. People's expectations for flexible and useable systems have risen greatly. In addition, the scope of new systems has grown considerably. In response, many designers have turned to usability as a means to ensure users (and, thus, organizations) can derive value from an ICT-based system.

Usability[1] testing was once the sole domain of European software designers. Today, however, there are professional organizations in the U.S. focused on usability approaches to software design (e.g., Usability Professionals' Association, see http://www.upassoc.org). In the U.S., the incentives to focus on usability are still primarily economic. For example, some firms that produce and distribute mass-market software, such as IBM and Microsoft, have invested in usability testing and problem identification programs in an effort to ameliorate customers' difficulties with their products. Today, most major software firms have their own usability labs. However, there is still a substantial gap between the complexity and prevalence of usability testing by these software vendors versus those who develop customized systems.

Although usability testing is becoming more common among mass-market software developers, these approaches are diffusing rather slowly into organizations that contract for customized ICTs for their own use. Most professional and educational literature still defines user involvement as assessing user requirements for a system only at the beginning of the design process. However, in these early assessments many users emphasize the major functions and routines of their work, overlooking important variations or exceptions. If user feedback is not continuously sought throughout the design process, then a new system is not likely to handle effectively overlooked exceptions, complexities, and nuances. In fact, most software development textbooks in the U.S. rarely discuss in much detail the social and technical complexities of designing useable ICTs (Salzman & Rosenthal, 1994; Truex, Baskerville, & Klein, 1999). We examine some of these issues in Chapter 5, "Teaching Key Ideas of Social Informatics."

Researchers who have studied the interactions between system designers and their clients have often found that designers rarely appreciate the work and working conditions of the people who will use the systems that they design (See Forsythe, 1992, 1994; Suchman, 1996; Ackerman, 2000). When computer systems are ineffective or fail, designers often blame poorly defined specifications (e.g., people who will use an ICT did not clearly express the problem) or technical limitations (e.g., hardware constraints). That is, although usability may reduce the problems with navigating a computer system's interface and may provide some guidance to designers, it is not likely to provide a means to address the issues of usefulness. By usefulness we mean the ability of the computer system to meet the needs of the users for whom it was designed (Woods, 2001).

3.2 Principles for Social Design

Here, we review the principles underlying the social design of ICT-based systems and in doing so discuss the concept of multiple users (and thus multiple needs), note that design continues after deployment, and speak to the issues of agency and users in design.

3.2.1 Social Design Compared to Designer-Focused Approaches

A growing number of researchers and practitioners are working to develop methodologies that take into account people's everyday work practices and activities (e.g., Davenport 1996; Denning & Dargan, 1994; Sachs, 1995). Davenport (1996) believes that system developers need specialists in the field (called "social systems analysts") who will be responsible for communicating with technologists and managing organizational change efforts. Tasks for the social systems analyst might include the following:

- "Shadowing" managers and workers to determine likely uses of the planned system

- Participating in system design efforts to ensure that the system fits the organizational structure and culture

- Facilitating user participation in the design activity

- Assessing current work practices and creating new ones

- Planning the implementation, including education and training

- Observing the system in use and making appropriate changes

Sachs (1995) exemplifies what a social systems analyst can do in her studies of one telecommunications company's uses of ICTs. Trained as an anthropologist, she is able to explore the social perspective of ICTs in a more nuanced way than would a person trained as a designer.

Increasing the number, visibility, and power of social systems analysts is important to better design ICTs for use. It is also likely to be contested because it does not fit the current designer-centric approaches espoused by most software engineering programs. That is, it is apparent that designers of software should be skilled observers of everyday work practices and activities in which a particular community of people engage (Brooks, 1996; Denning & Dargan, 1994; Sachs, 1995). There are some examples of practicing system developers who carefully observe people at work and engage them in design issues. For example, Gibbs (1997) reports how Mauro Mauro Design Inc. improved the performance of the New York Stock Exchange trading systems by improving the ways that traders could interact with four new systems. Their process involved six months of observing traders prior to coding new software, and 30 iterations in testing their new systems. Despite their behaviorally intensive approach, Mauro Mauro reduced the previous system development cycle from six years to two years.

The literature on systems design methodologies uses a variety of terms that embody the concept of socio-technical design. These include user-centered design, customer-centered design, action-centered design, participatory design, contextual design, cooperative design, usability testing, joint application design, and soft systems methodology. However, as Kling and Star (1998) point out, these terms "can easily become a trivialized buzzword that could casually be slapped as a label onto any computer application that seemed to help people."

As a way of defining what constitutes the design or use of human-centered computing, Kling and Star (1998) identified four aspects of the approach:

1. An analysis that encompasses the complexity of social organization and the technical state of the art. An analysis must consider various social units that structure work and information, such as organizations and teams or communities, as well as their distinctive social processes and practices.

2. A process approach that takes into account how criteria of evaluation are generated and applied, and for whose benefit. This should include the participation of stakeholder groups, such as involving patient groups in the development of specialist medical technologies, or teachers in the development of instructional technology.

3. An emphasis on more than just the formal functions of an ICT. As with the architecture of buildings, the architecture of ICTs raises questions of their livability, usability, and sustainability.

4. As part of both ICT design and evaluation, explicitly asking whose purposes are best served in the system's development. Through this questioning and the responses, the design rationales and decisions can be tied to the sources of influences. This knowledge of design is the essence of human-centered systems.

We use the term "social design of computerized systems," or "social design," to characterize the joint design of the technological characteristics of a computerized system and the social arrangements under which it will be used (Kling and Jewett, 1994). These "social choices" are an integral component of computerization, even though they usually are not formally decided or completely within the control of any one person. They can even be byproducts of oversight, such as managers who neglect to train their staffs in a new computing application because they assume that it's "just like the old one" or "very user friendly." An example of social choices can be seen in the example of phone line installation system detailed in Sachs (1995) and reviewed later in this chapter. The ICTs to support this kind of service are designed with assumptions about the number of technicians who will work on a service request—either one or a team. Choices such as these are social—they reflect the ways that work will be (re)structured.

As another example, organizations that adopt mobile computers to improve the flexibility of people's work situations and relationships must do more than simply acquire technologies to realize these specific advantages. Some of the desired convenience will hinge on technological choices, such as acquiring machines that run software compatible with that which is used in the office or that access high-bandwidth telecommunications networks. But much of the resulting flexibility will depend upon social choices as well. For example, if organization A requires its employees to report to work daily during regular working hours even when they have the portable computers, then people gain relatively little flexibility in work location if that policy is not also changed. People may be able to take mobile computers home after hours or on weekends, or on the road while they travel. Thus, organization A may gain some unpaid overtime work with these policies. But the men and women who work in organization A will not gain much of the "control over the location of work" which many people attribute to mobile computing. In contrast, if organization B allows its employees to work at places and times that they choose, then its employees can use mobile computing to have even greater flexibility in their work practices. In each of these cases, those who procured the equipment and constructed the work policies and practices for organizations *A* and *B* created distinct socio-technical configurations—or social designs. And these social designs, rather than the technology alone, will have different consequences.

Table 3.1 highlights the differences between the social design of ICTs and designs rooted in developer-centered engineering approaches. For example, engineering-centered conceptions of work and design take an explicit view, breaking activities down into defined tasks and operations. In contrast, socio-technical conceptions of work take a tacit view, looking at everyday work activities as a whole. Looking at activities as distinct from particular tasks means looking at how working people communicate, think through problems, forge alliances, and learn (Brown & Duguid, 2000).

The use of computer-aided software engineering (CASE) tools illustrates the differences in these two perspectives on design. CASE tools are designed to support software developers, and the primary focus of these tools is to guide designers to do certain production-focused activities. Empirical findings show that, instead, developers use these tools to support their social interactions. Guinan, Cooprider, and Sawyer (1997) and Vessey and Sravanapudi (1995) found that developers made minimal uses of CASE tools to support production, focusing their uses on communicating with one another.

3.2.2 Designing for a Heterogeneity of Uses, People, Contexts, and Data

A significant body of organizational informatics research emphasizes the importance of "user involvement" in the design of new systems. By the 1970s, research monographs such as Henry Lucas's (1975) *Why Information Systems Fail*, and analytical reviews of the studies of systems failures (e.g., Kling, 1977) identified as one major cause of system failures the exclusion of the people who will be using the system from the design process. Such failures are reported in the Human–Computer Interaction research literature (Poltrock & Grudin, 1994; Wilson, Bekker, Johnson, & Johnson, 1997) as well as the information systems (Montealegre & Keil, 2000) and computer science literatures (Clement, 1994). In addition, industry surveys show that as many as 40% of systems projects in major corporations are total failures and that not having an ICT's actual users participate effectively in the design of the systems is one major cause (James, 1997).

It is not hard to find professional articles that echo these research findings with calls for designers to "involve the people who will use an ICT" or to build "user-friendly" systems to ensure their usefulness. As we have already noted,

Table 3.1 Approaches to Designing ICTs for Workplaces

Designer-Centric View	Social Design View
Explicit views of work	**Tacit view of work**
Work can be documented, made visible, and thus easy to articulate and transfer	Important aspects of work are silent, shared, and come to be understood by the workers
Training makes work possible	Learning makes work possible
Tasks are the core of work	Knowledge is the core of work
Position is made clear in the hierarchy	Informal political networks and networks of contacts define position
Procedures and techniques are basis of doing	Conceptual understanding is the basis of doing
Work flow represents what is done	Work practices represent what are done
Methods and procedures are the guides	Rules of thumb and judgment are the guides
Intended Goals	**Intended Goals**
Improve efficiency	Improve work practices
Reduce human error	Help people discover and solve problems
Design Assumptions	**Design Assumptions**
User needs are identified by what is visible and documented. These can be rationalized into one set of needs.	Users needs emerge from observing everyday work practices. These often conflict and there are often real differences in needs.
Design is linear and can be documented at the end	Design is iterative and requires prototyping
Individual work is to be supported through process clarity	Collaboration and collaborative learning take place in a social context
Efficiency is a desirable outcome	Skill development is a desirable outcome
Technological Choices	**Technological Choices**
People can adapt to technologies chosen to support organizational values	Configurations matter and interact with human activities, such as work

usability labs can be helpful for refining some aspects of an interface's design. However they don't help designers learn how a person will use a system in concert with other technologies (such as other systems, reports, books, telephone, fax, etc.) in their own workplaces. Many people think of improving usability in terms of improving what a user sees on an ICT interface (such as menus, buttons, etc.). Although this focus can yield many improvements, it is too narrow; there are many usability issues that are "more than screen deep" (King, 1999b).

Furthermore, ICT design communities are not monolithic. Grudin (1991) distinguished between three major kinds of design contexts. In *contract development*, the customer, or user organization, is known from the outset, and the development organization is identified when a contract is awarded. In large contracts, few designers actually visit the user organization. In *product development*, (e.g., "shrinkwrap" products), the developers are known from the outset, but the people who will use an ICT typically remain unknown until the product is marketed. Finally, in *in-house development*, both the people who will eventually use an ICT and its designers are known at a project's outset. (This is also known as *custom development*, where a specific external developer is engaged from the start to produce or configure a system for a specific customer.) These distinctions help us understand the different conditions under which ICT designers learn about the work and settings for which their systems are intended. For example, in-house development of ICTs makes it easier to establish and maintain continual contact with likely users. In contrast, contract developers may work thousands of miles away from the people who will use their systems (Grudin, 1991).

When designers are working without much contact with their intended clientele, they frequently develop abstract categories of a system's potential "users." However, the linguistic convenience of easily labeling the people who might use an ICT masks their likely diversity. The diverse groups of people who use an ICT, and their uses of it, are not best understood by simple categories such as "novice," "expert," or "casual." Further, many designers develop tacit scenarios of the ways that people will use systems that often differ from many actual conditions and uses.

Oversimplified scenarios of an ICTs use can have serious costs. An extreme example illustrates how the designer's scenarios of use may differ substantially from actual working conditions. In July 1988, the USS *Vincennes*, patrolling the restricted waters of the Persian Gulf, shot down an Iranian Airbus 320 civilian airliner with 290 civilians on board. This extraordinary and tragic event led to formal inquiries by the U.S. Navy and the U.S. Congress, as well as an inquiry by five highly qualified psychologists. It was also investigated by computer science and political science researchers, whose studies form the basis of our account here (e.g., Rochlin, 1997, Chapter 9). Significantly, researchers found that the *Vincennes'* RCA-designed Aegis missile defense system performed correctly at the time that the Airbus 320 was shot down.

The events of that day are complex, but included an attack on the *Vincennes* by several small gunboats, during the course of which the captain turned the ship rapidly, resulting in the ship's tilting at a steep angle during this surface battle. According to Rochlin, "the effect was particularly dramatic. Books, publications, and loose equipment went flying off desks. Desk and file drawers flew open. Many of those on duty had to grab for the nearest support to avoid being thrown to the deck....The situation aboard the *Vincennes* that day was one of confusion and disorder."

Almost simultaneously with the surface attack, the *Vincennes'* crew interpreted their Aegis displays as signaling an attack by an incoming F-14 fighter. The operator believed that the airplane was descending, although replay of the data clearly indicated an ascending civilian plane. However, once the crew developed the F-14 scenario, they remained committed to it, acted in concert with it, and after ineffective efforts to communicate with the F-14 on military voice channels, were authorized to fire two missiles in order to bring it down.

The Aegis system aboard the *Vincennes* had been designed by drawing on findings from a range of scenarios. However, no scenarios covered the chaos and complexities that the *Vincennes'* crew faced that day. Matt Jaffe, one of the designers of the Aegis display interface, reported that the altitude information was difficult to interpret correctly. Thus, it could become difficult to determine whether the plane was going up, or going down, or remaining at the same altitude (Neumann, 1995). These displays were probably adequate under the specific scenarios envisioned by the designers. However, the designers probably never considered that civilian aircraft, rather than "friendly" or "hostile" combat aircraft, would be so close to a ship like the *Vincennes* under battle conditions. Here the system's scenario-based design both increased the value of the system during operations and, at the same time, made it difficult for its users to adapt to unexpected situations.

The U.S. Navy was deeply invested in Aegis, and often officially defended it. The U.S. Navy's report on this tragic mishap blamed combat-induced stress as a major factor, thus placing the burden of the problem on "incapable users" while simultaneously exonerating them with a claim of battle stress. The psychologists, who did not have a stake in defending Aegis, felt that the Navy's analysis did not effectively engage important factors. The psychologists criticized the Navy's report and testified to the U.S. Congress that "operator-error" was no longer a suitable explanation for mishaps with complex weapons systems. Rochlin comments:

What are the expectations of a combat system such as the Vincennes, or of the CIC as a war-fighting center? That stress would be low? That battle conditions would be other than confusing? That the ship could be attacked on the surface, or from the air, but not both simultaneously—not to mention possible subsurface attacks in other circumstances? If these are or were the assumptions under which the Aegis cruisers were designed, than the Vincennes should never have been deployed into the Gulf.

The tragic shooting of Iran Air Flight 655 was an extraordinary event whose circumstances were not effectively anticipated. RCA's design engineers, interface experts, and combat-experienced naval personnel, working on an arms-length contract during peacetime, may have had serious problems in foreseeing the battle conditions under which Aegis would be most important. However, there are also many cases in which system designers have ample opportunities to evaluate the conditions under which people will use systems, and still fail to do so.

In a less sensational, but no less evocative example, consider the issues with educators' uses of the Web. In one department of a leading university, faculty posted their syllabi on Web pages. Links to the pages, and often the pages themselves, were changed from one term to another. This maintenance of pages and links constituted a burden for the department's technical support staff. Without consulting the faculty, they designed a Web-accessible database to maintain all of the syllabi. They portrayed the shift as a technical streamlining. In their view, "the users" of the syllabi collection were the department's faculty and students who would know the name and number of the course whose syllabus they are seeking. But other people unfamiliar with the course titles and course numbers, and who were invisible to the department's support staff, used the syllabi as well. These included prospective students and faculty who were based in other universities and who were developing courses. The shift to a database also hides the syllabi from search engines, and thus from faculty elsewhere who are searching for sample syllabi and teaching materials.

This technological shift may have simplified the technical staff's ability to maintain the department's Web site, but did so with some losses to the field and to the school's external visibility. Again, an ICT-based system's design both constrains and enables—this is a given. The ways in which constraints are made, and enablers provided, is tied to their use. Thus, design becomes a negotiated principle (Truex, Baskerville, & Klein, 1999; Truex, 2001).

3.2.3 The Designing of ICTs Continues During Their Use

The computerization of common organizational activities, such as accounting, inventory control, or sales tracking, is not a one-shot venture. Computerized systems that are introduced at one time are often refined over a period of years (Kling & Iacono, 1984) and are periodically replaced by newer systems. Some computerized accounting systems have histories of 30 or 40 years (Copeland, Mason, & McKenney, 1995), and 10 to 20 years is quite common in manufacturing.

These examples, plus evidence from empirical studies of actual work practices (Greenbaum & Kyng, 1991; Johnson & Rice, 1987; Kraut, Dumais, & Koch, 1989; Suchman, 1996; Wagner, 1993; Walsham, 1998), reveal that the people and groups that utilize ICTs reshape them in ways that their original designers did not anticipate. This reshaping of system use suggests that an ICT will change from its initial design. Greenbaum and Kyng (1991) identify three reasons why systems design does not end with the delivery of a final product or service:

1. Because most ICTs are designed to be used for long periods of time, circumstances or the situation of use are likely to change (i.e., needs change, uses change, people who use the ICT change, and the organization changes).

2. The complexity of systems design and the context of use makes it difficult, if not impossible, to anticipate all the issues that will eventually be of importance in the final design. It is inevitable that the designers of ICTs and the people who use them will overlook important issues.

3. The flexible use of ICTs by different groups of people requires that ICTs be designed for many different situations of use. Sophisticated software "packages" are often designed to have multiple configurations to allow them to satisfy as many users as possible with a single product.

Configuring a system, or continuing design into and through use, is an activity different from initial design. As Greenbaum and Kyng (1991, p. 223) observe: "The activity is related to specific use situations and the result is not a new system, but a modified system; that is, a system with a history which relates it to the earlier version and problems with its use." Many instances of system failures might not have had such adverse outcomes if the developers had not left the project on its "delivery date." For example, the designers of the Aegis interface, described earlier, may have improved the clarity of the altitude readings if they had realized earlier the limitations of the combat system in a hostile environment.

In addition to understanding specific use situations, good design often requires a critical, use-oriented perspective to help ensure that unintended problems/losses do not result. It is very rare for designers to develop highly successful systems without a substantial understanding of the conditions under which people will use them. Unfortunately, many systems developers and consultants who focus on organizational goals fail to acknowledge how work is carried out in practice. The following case illustrates this point.

One major phone company made a business of installing high-speed phone lines in large offices. These installations were not being completed in a timely manner, and the company began to lose customers to competitors because of the delays. Consultants hired to help improve efficiency observed technicians talking with each other and sometimes waiting for each other so they could go out jointly to a customer's site. The consultants recommended that the conversations, which were viewed as socializing rather than working, be greatly reduced, and that technicians be sent out on jobs as soon as they were available. In their view, any technician who had received certain standardized training should be able to fulfill any service request.

The consultants developed the Trouble Ticketing System (TTS), a large database that also functioned as a scheduling, work routing, and record keeping system[2]. Although TTS was designed to make work more efficient, it actually had the opposite effect and failed to achieve the company's efficiency goals, because the consultants had not accounted for several important aspects of the technicians' work practices. One of these aspects was the technicians' conversations with each other, which TTS was designed to eliminate:

> In these conversations, they compared notes about what was going on at each end of the circuit. If there was a problem, they figured out what it was and worked on it together. These trouble-shooting conversations provided the occasion for workers to understand what was actually going on in the job, diagnose the situation, and remedy it (Sachs, 1995, p. 39).

The consultants also failed to recognize that the technicians had developed specializations after finishing their common training, and that some were thus better able to handle different types of problems. Further, although TTS did not allow for this, the technicians often needed to work jointly with each other in order to deal with complex situations.

When social design is preempted by designer-focused approaches, the resulting ICT-based systems often ignore actual work practices and the users often will (or must) develop *workarounds* as a way of dealing with poorly designed systems. In the case of TTS, for example, workers soon figured out new ways to contact appropriate coworkers in order to problem solve. They then used TTS to provide a record of the conversations by misrepresenting the conversations to appear as though they were work done on site (Sachs, 1995). However, despite the workarounds, the phone company's installation process did not speed up to the extent predicted by the consultants, and they were not regaining their lost market share. In desperation, the company hired an anthropologist to study how the work was really done. With the cooperation of the technicians, the new consultant developed a team-based approach to high-speed phone installation. This new approach was substantially more effective in increasing the speed of the installations and allowing the company to regain its market share.

3.2.4 There Is Agency in the Design and Deployment of ICTs

Stakeholders for an ICT design differ, depending on the nature of the ICT and the designing organizations. The stakeholders may include the design team itself, the people who will actually utilize the ICT, people and groups that depend upon the functioning of the ICT (even if they do not use it directly), and the designing organizations. Thus, the development of ICT applications requires the collaboration or involvement of a variety of distinct communities, composed of workers with different skills using different representational frameworks. This necessary heterogeneity poses a number of problems that cannot be removed simply by ensuring good communication between the differing groups.

For example, the Worm Community System was an information system for helping molecular biologists who worked in hundreds of university laboratories to share information about the genetics of certain worms (nematodes). It was designed as a UNIX-based system because its designers felt this was the most beneficial technical environment. In use, it required a sociotechnical infrastructure comprising network connectivity and UNIX computing skills. These skills had to be a part of each laboratory's (user) work organization—and also the local university's resources (Star & Ruhleder, 1996). Star and Ruhleder found that the Worm Community System was technically well conceived. But it was actually rather weak in supporting scientific communication because of the uneven and often limited support for its technical requirements in the different university labs. The system had been designed by a group of computer scientists who preferred UNIX; in making this decision, they failed to realize the technical requirements they placed on their users. Few of the bio labs that could benefit from the system had people with UNIX expertise, and therefore they found the system puzzling and cumbersome to work with. In short, a lack of attention to the local infrastructure (of the users) can undermine the workability of larger scale projects.

Groups or organizations that commission a customized ICT (such as a state motor vehicle agency seeking a newer vehicle registration system) have to mobilize and coordinate the resources to design an ICT package (or subcontract the project). This mobilization effort is rarely a neutral activity; it is easiest for upper managers or entrepreneurs to mobilize resources by emphasizing the social or organizational values that a new ICT can enhance. However, there may be a conflict between the values of those who want to have a given system used and those who will have to use it. For example, the manufacturing division of one major computer company developed an elaborate order entry system to schedule orders. In order to generate a price for customers, the salesmen, who worked in a separate sales division, had to specify *all* of the computer's components, from the expensive hardware down to the relatively inexpensive connecting cables. Manufacturing staff wanted this type of detailed and accurate specification, because their division bore the costs of configuration errors or "dirty orders." However, many salesmen wanted to use the system to give customers rough estimates of the prices of different configurations. The salesmen were rewarded for their sales volumes and were not penalized for "dirty orders." Thus, the salesmen ended up rarely using the system because it required too much time and effort to fill in all the information about the numerous inexpensive parts that the system required them to specify. This system was not able to fulfill the needs of the salesmen and their customers for quick estimates.

Manufacturing managers and their system analysts thought that a new interface would make the system more usable for the salesmen. They conducted a multimillion dollar interface redesign project, which had little effect. Although the interface was improved, the redesign did not address the conflicts of values between the manufacturing staff and the sales force.

Similarly, assumptions about how people work are often embodied in ICT design choices (Agre & Schuler, 1997). For example, e-mail systems that sort messages for a person to read based on technical criteria such as their recency or length also influence the recipients' social relationships by encouraging more attention to some senders and their message than to others. An analysis that informs design should not focus only on optimizing the technical capacities of ICTs, but also on recognizing and supporting the social relationships that they influence (Kling & Star, 1998).

Technological artifacts may embody a single set or multiple sets of values that may influence the design of ICTs. In these complex design contexts, representing multiple views is difficult, messy, and often politically charged (Bowker & Star, 1999, p. 41). The design of ICTs that work well for people and help support their work, rather than make it a little more complicated by getting in the way at times, is a subtle craft. It is not obvious, and it cannot be well taught without an understanding of how people work and what kind of organizational practices obtain. Simple criteria—such as get *more* advanced technology (whether it is faster or easier to use); get *better* technology; or organize systems so that they are more efficient—have not been good enough (Kling, 1999b). Organizations will have to learn to cope with issues such as re-evaluating how work is done; increasing communications between administration, technical support, and front-line employees; re-training employees; and continuously evaluating changing hardware and software needs. We should promote the design of ICTs that are really workable for people, rather than simply advocating technologies that may occasionally work and occasionally be valuable, but are often unusable, and thus incur extra cost and misplaced hope.

Endnotes

1. The term *usability* is used to characterize the extent to which a specific ICT is relatively easy for a person to utilize for specific activities. Usability differs from *beta testing* in two ways. First, beta testing focuses on the overall operation of new code (such as the level of defects). Second, it is a chance to gauge user interest (for purchase or adoption). Usability tests typically focus on direct user interaction with the screen and the system's features.
2. TTS is discussed in Sachs (1995).

CHAPTER 4

Social Informatics for ICT Policy Analysts

> The information revolution ... has evolved at a different pace than suggested by the hype surrounding it, and it has taken many unanticipated turns. For example, the convergence of different modes of communication (voice, data, audio, and video) that has been proclaimed for decades only now is happening, slowly and in forms that defy earlier visions. During the past two decades, an increasing array of telecommunications policies was designed to allow competitive market forces to flourish, yet the main players of a quarter century ago still dominate the industry. On the other hand, the rapid growth of wireless communications or the Internet was not foreseen.
>
> (Bauer, 2001, p. 413)

> Communications policy is characterized by constant struggles and disagreement and not by some monolithic ideology. Policy "issues" are always hotly contested. Knowing what those issues are, who is involved, and what is at stake is part of the job of the communications policy expert. Communications policy is hardly a straightforward matter of engineering. It is not a clean, neutral, predictable, mechanical, or routine process.
>
> (Streeter, 1996)

This chapter examines how Social Informatics (SI) research can be of value to information and communication technology (ICT) policy analysts. According to Weingarten (1996; p. 45, 46), information policy is "the set of rules, formal and informal, that directly restrict, encourage, or otherwise shape flows of information ... [it is] a balance between competing social interests; and information technology affects those interests through what people and institutions do with it." ICT policy analysis is concerned with determining the extent to which governments and public agencies should become involved in promoting the use and extending the reach of ICTs in society. A wide array of public policy issues fall within the ICT policy analyst's domain, from the reduction of the digital divide, to the promotion of electronic business, to changes in corporate media control, to the protection of

51

digital intellectual property on computer networks. As analysts investigate these issues, their primary goal is to influence the beliefs of policy makers by clarifying the issues and defining and evaluating the alternative positions that can be taken with respect to these issues (Thorngate, 2001, p. 85).

As Streeter points out, technology policy analysis is a particularly difficult and sometimes contentious activity because of the high stakes and range of powerful stakeholders who attempt to influence policy makers' decisions about the management of their governments' financial, legislative, and regulatory involvements in ICT development and use. Bauer (2001, p. 413) explains that ICT policy analysis is an activity fraught with complexity because these technologies do not interact with "established social and political processes" in stable or predictable ways. In fact, "more often than not, new technologies actually create many new problems, fall far short of their predicted abilities, or bring with them a myriad of unintended consequences" (Cranor and Greenstein, 2003, p. 1). One implication of this situation is that the evaluation of potential outcomes of competing policy alternatives becomes difficult for the analyst, introducing a level of uncertainty into the process that is accelerated by the dynamism of technological change. The main question raised in this chapter concerns the ways in which SI research can help to mitigate this uncertainty, thereby improving the policy making process.

4.1 How Can Social Informatics Contribute to ICT Policy Analysts' Work?

In this section we explore the notion that some of the findings and theories of Social Informatics research can inform ICT policy analysis, even when the research does not evaluate specific policies. Research examining sociotechnical assumptions that underlie thinking about ICTs, but that do not directly impact specific policies, is often called policy-relevant research. Porter and Hicks (2000) observe that policy-relevant research can be very influential in policy analysis and policy making, but that it:

> ... seldom has an immediate or direct impact on government decisions; more typically its influence is indirect and incremental. New technical knowledge tends to 'creep' into policy making, gradually altering the background assumptions and concepts that frame policy discourse. Because these kinds of cognitive and linguistic shifts are fairly subtle and diffuse, and often are only observable over a span of years, the impact of (social) scientific research and analysis on policy making is often underestimated.

In this section, we explore a range of ways in which SI research can help ICT policy analysts in their work. In Section 4.1.2.1., we examine an example of technological innovation and change, the ICT policies that were developed to accommodate these changes, and the contributions that SI could make to the resulting policy debate.

4.1.1 How Social Informatics Can Help

Policy analysis is not a linear process. It is not a rational sequence of discrete steps that result in the formulation of policy alternatives. Research indicates that it is actually "very untidy, with outcomes occurring as a result of very complicated political, social and institutional processes, which are rarely obvious," (Fitzduff, 2000, p. 3). Most policy analyses rest, in part, on the analyst's training, prior experiences, social philosophy, institutional affiliations, and applicable legal and regulatory regimes. However, ICT policy analyses also rest on assumptions about technology. Beliefs about how ICTs are developed, how they are used in practice, and the range of social consequences likely to accrue from their use all play a role in shaping the analyst's view of a given technology policy issue. SI research can be used to challenge and transform these often taken for granted assumptions that policy analysts have about the nature of technology and its impacts on social and work life. For instance, a technologically deterministic approach to defining ICT policy would assume that social changes could be brought about simply through the introduction of the appropriate technologies in the appropriate places at the appropriate times. An approach steeped in SI might emphasize three themes, which were highlighted in earlier chapters, embeddedness, configurability, and duality of ICTs, providing a richer and more nuanced basis for analysis (see Figure 4.1, Key Social Informatics Themes).

Pal (quoted in Abrams, 2002, p. 14) defines policy analysis as "the use of multiple methods of inquiry, in the context of argumentation and debate to create, critically assess, and communicate policy relevant knowledge." Given this definition as a starting point, we believe that ICT policy analysts can benefit from SI theory and research in several ways. SI is one of the appropriate methods of inquiry that should be used because of its focus on the relationships among ICTs; the people who design, use, and manage them; and the contexts in which they are used. At this more abstract level, SI can provide policy analysts with a set of reliable and tested conceptual frameworks that they can use to isolate, organize, and understand the social, cultural, and organizational forces affecting ICTs and their uses in ways that enhance the development of policy-relevant knowledge. It can provide them with a set of reliable analytic techniques to help identify and evaluate the social consequences of the implementation and uses of ICT-based systems. It can help

Embeddedness: ICTs do not exist in social or technological isolation—their cultural and institutional contexts influence their development, implementation, use, and role in organizational and social change.
Configuration: ICTs are socio-technical networks that can be configured in ways that influence their uses and social consequences.
Duality: ICTs have both enabling and constraining effects on groups, organizations, and larger scale social orders.

Figure 4.1 Key Social Informatics Themes

them develop a more critical appreciation (by which we mean reflective, inquiry-focused, and problem-based) of the benefits and limitations that ICTs can provide in social and organizational settings. Finally, SI can help them understand how the design, configuration, and implementation of ICTs is a socio-technical process—one that is charged with political, economic, social, and organizational implications.

ICT policy analysts must often center their analysis on specific technologies, such as cell phones, broadband, or wireless communication. What we have learned from SI is that it is helpful for analytic purposes to move beyond the conceptualization of ICTs as discrete objects and view them instead as configurable "socio-technical networks" made up of tangible and intangible components including:

- People in various roles and relationships with each other and with other system elements (designers, content providers, managers, users)

- Hardware (computer mainframes, workstations, peripherals, telecommunications equipment)

- Software (operating systems, utilities, and application programs)

- Techniques (management science models, patterns of use and configuration)

- Support resources (training, support, help)

- Information structures (content, rules, norms, and regulations, such as those that authorize people to use systems and information in specific ways, access controls, security measures)

These components are not simply a static list, but are interrelated within a matrix of social and technical interdependencies. As has been explained in earlier chapters, this socio-technical network model has been useful for understanding ICTs and the people who design, implement, and use them in a wide variety of organizational and social settings. It has substantial repercussions for ICT policy analysts who seek to understand how ICTs are actually used in social and organizational settings because it focuses on the web of technology and not just on the technology itself. With this type of approach, there is a range of relevant insights that can be incorporated into ICT policy work.

ICT policy analysts can use SI research to understand how and why the design, development, implementation, and uses of ICTs are not value neutral. In fact, these are distinctly socio-technical processes that are heavily influenced by the approaches and orientations of stakeholders who have varying degrees of interest in ICTs' development, deployment, and use. SI research has shown that there are typically positive and negative consequences from the implementation of ICTs in organizational and social settings. In general, once in place, ICTs enable and constrain social actions and social relationships and affect workflow and other organizational processes. Analysts can use these findings to assess the extent to which different ICT policy decisions will, if carried out, lead to advantages for some and disadvantages for others; some groups will gain in power, prestige, authority, and control of resources, while the relative status of other groups will be diminished.

4.1.2 Contemporary Policy Issues from a Social Informatics Perspective

What follows is a case study that illustrates the extent to which SI issues are central to many ICT policy issues. The case demonstrates how ICT policy analysts might have used an SI approach to modify, rethink, or abandon an ICT deployment proposal. We conclude by offering suggestions for how ICT policy analysts can incorporate SI concepts, research findings, and themes into their current and future discourses.

4.1.2.1 Notebook Computers Replacing Textbooks

During the latter half of the 1990s and in the first years of this decade, many K–12 school districts and corporations in the United States explored the potential of using notebook or laptop computers in educational practice. According to Keefe and Zucker (2003, p. 9), "large school districts (e.g., Henrico County, VA) and even whole states (e.g., Maine) are adopting initiatives involving laptops using wireless networks connected to the Internet." Apple, Microsoft, Dell, Toshiba, IBM, and other vendors are wholeheartedly supporting these initiatives with grants, equipment, and in-kind donations. Mobile computing for students seems to offer a solution to a range of problems that were apparent in many locations such as the high cost of textbooks, the decreasing budgets for school libraries, and the subsequent lack of print resources for students. Some commentators are very optimistic about the future of notebook computing in K–12 education. McKenzie (2001) argues that "wireless notebook computers may prove an ally in the effort to recruit the enthusiastic daily use of those teachers who have been hitherto reluctant, skeptical and late adopting when it comes to new technologies." However, according to Belanger (2000):

> The future of mobile computing in K–12 education is still uncertain. Laptops may never become as common in classrooms as hand-held calculators. Solutions for issues of cost, technical support needs, security, and equitable access are challenging for many schools. Many schools with laptops, however, remain positive and enthusiastic about the changes observed and benefits their students derive from access to portable computers.

Can Social Informatics help the policy analyst evaluate this uncertain future? It is instructive to look at a policy debate that took place in the late 1990s in Texas over the introduction of notebook computers into K–12 education because it illustrates the ways in which an SI-informed policy analysis could have positively influenced the debate. In the fall of 1997, the Texas Board of Education proposed what seemed to be a very cutting-edge idea: replacing public school textbooks with notebook computers, one for each student. Facing a $1.8 billion bill for the purchase of new textbooks over a six-year period, the Board's motivation seemed to be to cut costs. The Board developed a plan to put durable, low-cost notebook computers into the hands of every Texas public school student, estimating that this would save the state nearly $300 million each year. These projected savings would come from shifting materials costs from the state to parents. Instead of purchasing textbooks, the state would require parents to lease notebook computers for their school-aged children. The monthly lease fee would be $10 per child per month and the total lease cost would be $500. At that time, the state of Texas was spending $450 on textbooks for each student per year, or $37.50/month (Christie, 1998). Asking parents to pay for books or other educational materials is not unusual. Indiana, for example, charges parents an annual $100 textbook rental fee for every child they send to public school.

The plan was presented to the state legislature in May 1998. Surrounded by supporters from several educational software vendors, proponents of the plan predicted higher test scores; happier, more engaged students; future tech-savvy employees; and significant cost savings. They argued that notebook computers would be better than textbooks because available educational software would be more current (as well as inexpensive and quick to update), and the notebook computers would contain modems so that students could connect to the Internet. With the state saving so much money, some supporters envisioned additional benefits, such as a reallocation of funds to construct new schools in order to decrease classroom size and teacher–student ratios.

Despite all the possible economic and educational benefits that could accrue from the statewide adoption of laptop computers in Texas schools, the proposal did not receive legislative approval. Why would such a pro-education, pro-student proposal stall in the legislature? While the plan seemed to have key ingredients to ensure easy approval, questions such as the following may have caused some legislators to scrutinize the proposal more closely:

- To what extent do notebook computers enhance student learning in K–12 schools?
- To what extent do they increase the value of education?
- What are the actual costs of supporting millions of notebook computers for students and teachers?

With regard to the first question, some other states have used ICTs to attempt to remediate some public education deficiencies. School districts in South Carolina, Ohio, and New York have been experimenting with variations of "the notebook computers for students program" for some time, albeit with mixed results. A teacher in one poor New York City school district claimed that notebook computers made one of her students "write longer and more thoughtful assignments" (Stoll, 1998). Other teachers have been more skeptical about the benefits of replacing books with notebook computers and argue there is no systematic research that demonstrates that computer use improves student learning or achievement. In fact, educational researchers grounded in curriculum and instruction theory should conduct more empirically anchored research on the role that computers play in K–12 classrooms (e.g., Coley, Cradler, & Engel, 1997; Schofield, 1995).

The answer to the second question, if based on assumptions of technological determinism, is easy. The introduction of this technology should positively impact student learning in measurable ways, increasing the overall value of education. A more nuanced analysis based on an SI approach would find that the technology (notebook computers) itself is embedded in a complex and dynamic social and organizational context of students, parents, support personnel, schools, regulatory agencies such as State Departments of Education, parent–teacher groups, and a range of social practices (such as teacher's work flow and a student's study habits). Using this approach, a policy analyst would look to the SI research that examines the use of notebook computers in educational settings taking this larger web of social and technological relationships into account (see, for example, Pew Internet Life Project, 2002). The subsequent analysis of policy alternatives can then incorporate these findings and provide policy makers with a richer understanding of the possible consequences of their decisions.

The third question, that of technical support infrastructure costs, is more complex. The Texas proposal focused on cost savings. However, it did not provide a short or long-term plan for ICT implementation, training, upgrading, and maintenance. An organizational informatics analysis of this proposal would question the absence of such plans. A smooth implementation of any networked system requires, at the very least, some technological standardization within school districts or at least school sites. Computer training requires a level of standardization and a significant time commitment, especially when working with children who have widely varying skills, literacy levels, abilities, and interests. There is also the question of the ways in which the curriculum would have to be changed to accommodate the introduction and widespread use of this technology. Can teachers be expected to know how to alter their lesson plans, classroom activities, and homework assignments to take advantage of these computers? None of the potential benefits comes without an additional price tag, and it is incumbent upon ICT policy analysts to understand the ripple effects of their recommendations on the larger society.

Organizational informatics (OI) can help ICT policy analysts understand the significant costs associated with providing a technical support infrastructure for the notebook computers in Texas (or in any state). This kind of analysis would consider small hidden costs that could scale up, such as the cost of spare notebooks to replace damaged units, spare batteries, power adapters, or even the increase in school electricity bills because notebook battery life is limited and students would have to have their notebook computers plugged into electrical outlets for the majority of the school day. It could also identify major expenditures that would come from having to hire technical support personnel to maintain and upgrade millions of notebook computers now owned by the school system and people to train those involved in the implementation and use of this technology.

Using an OI analysis to understand the implications of the Texas notebook computer proposal yields an interesting insight for the ICT policy analyst. In the proposal, the claim was made that the cost savings to the state would be an estimated $2 million. The current industry technical support model assumes that there is one systems administrator (who typically earns $50,000/year) for every 100 computers. A simple calculation using these and the numbers in the proposal shows that the monthly cost of providing a minimal technical support infrastructure would be approximately $42 per machine per month. One implication of this analysis is that the ability to pay must be factored into the evaluation of policy alternatives. It is not likely that all Texas parents could afford $52/month ($10 notebook lease fee plus $42 support fee) per student. How can these costs be covered for those who cannot afford to pay? If the state of Texas agreed to pay the $42 fee, the projected $1.8 million cost savings would quickly evaporate. Instead, there would be an increase of $54 per student. In short, an analysis informed by OI would help the ICT policy analyst understand that there are cost-based problems with the proposal. Such an analysis would have indicated that, in all likelihood, the proposal would not have been able to deliver the cost savings that had been promised.

Outside of the stipulation that the notebook computers would be equipped with modems, the initial vision of statewide notebook computing in primary education did not include a networking element. However, if the Texas proposal had followed the lead of the Clinton-Gore policy to extend Internet connectivity to all K–12 school in America, it would have included a plan for networking the notebook computers to allow access to the Internet. Social informaticians would also have examined issues of pedagogical value, Internet filtering, and intellectual property rights. It is worth commenting briefly on the first of these issues—pedagogy. Most of the rationales for providing Internet access to K–12 schools are based on (often tacit) assumptions that such access will enable students to learn through inquiries that are enriched by a wide array of resources that cannot be readily found in K–12 school libraries and locally in many communities. Most of the educational technology projects funded by the National Science Foundation in which schools use Internet access also emphasize inquiry-based pedagogies. This model has strong appeal to many academics who see processes of disciplined inquiry as an important element of continuous personal learning (as well as being reflective of a scientific or scholarly orientation, more generally). However, a new inquiry-oriented set of teaching practices that integrate large numbers of new resources, such as the heterogeneous array of reports and discussions available through the Internet, takes significant time for teachers to develop, evaluate, implement, and refine. Unfortunately, K–12 teachers have relatively little slack time for developing new teaching approaches.

The point of these observations is not to criticize the national project of connecting K–12 schools to the Internet. The point actually makes use of a deeper Social Informatics finding—that the value of ICTs usually comes when social practices are changed along with the technologies; changing (or adding) technologies alone rarely produces many transformative benefits. In the case of providing K–12 schools with Internet access, the Presidential Advisors note that school reform based on technological innovation and change will also require giving teachers more time to learn how to integrate digital networked resources into enriched forms of teaching.

Social and organizational informaticians have already researched infrastructure issues similar to the Texas notebook computer proposal (Gasser, 1986; Kling, 1992; Kling & Scaachi, 1982; Kling & Star, 1998; Schmidt & Bannon, 1992; Schofield, 1995; Star & Ruhleder, 1996; Suchman, 1996). Their research demonstrates that the deployment of ICTs has far-reaching but subtle consequences and indicates that those responsible for ICT deployment often overlook or fail to see those consequences. The Texas notebook computer proposal is a good case in point and one that could have benefited from SI research.

An SI approach to the analysis of notebook computers in primary and secondary education could also aid the ICT policy analyst by providing a clearer understanding of the web of complex social and organizational relationships in which the computers would be enmeshed. From our earlier discussion of the adoption of ICTs in organizations, we discussed how local incentive systems strongly shape behavior; in fact, they are often "designed" to do so. A closer look at the case of Texas, however, illustrates the difficulties of policy analysis when there are conflicting incentive systems. On one hand, there is a proposal to implement a fundamental technology-based school reform; on the other, there is an equally strong commitment to standardized testing. Texas is one state whose politicians and State Board of Education have embraced high-stakes standardized testing in an effort to improve education and the accountability of schools to their publics. High-stakes standardized testing—in which students' scores on a standardized test are used to decide such matters as who will receive high school diplomas, what the annual raises will be for teachers and school principals, and the kinds of state funding that will be available for school sites—is a highly controversial educational reform (Khattri, Reeve, & Adamson, 1997). Some educational researchers are enthusiastic about standardized testing, arguing that it helps to improve school accountability, but they are less enthusiastic about high-stakes testing. In contrast, others claim that high-stakes standardized testing reduces the quality of education for many students, because many teachers "teach to the test" (Grant, 2000; Mabry, 1999; Rallis & MacMullen, 2000) in ways that limit the effectiveness of the curriculum and de-emphasize "authentic" inquiry-oriented instruction.

Using an SI approach, a policy analyst might argue that high-stakes testing is intended to shape the behavior of school administrators and principals. The high-level rationale is that the testing regimes will lead teachers to teach in ways that improve students' learning. However, those teachers who spend their limited "slack time" learning the tests and how to teach test content, will have even less time to develop the inquiry-oriented pedagogies that notebook computers and Internet access could support. In short, the two reforms—expanding inquiry opportunities with Internet access in the K–12 schools and tightened school accountability through high-stakes standardized testing—clash because they impose conflicting demands on how teachers use their limited time. The extent of these conflicts in practice is an empirical matter. Indirect evidence indicates that the different educational reform efforts that are sweeping the U.S. may be incompatible in terms of classroom practices. This does not mean that reforms should not be advocated. Rather, it indicates that the concrete demands of reform programs in the classroom and on students' and teachers' time and orientation have to be brought center stage to set appropriate expectations for their practical value.

This example illustrates that the use of an SI approach could help ICT policy analysts in their work. By taking advantage of what SI researchers have learned about the embeddedness of ICTs in complex social, cultural, and organizational settings, analysts could more easily disentangle the notebook computer proposal and clarify the range of reasonable and appropriate policy alternatives. In doing so, they would take into account such socio-technical issues as the costs of the technology implementation for school systems, State government, and parents; the costs of providing adequate technical support, the development of training programs for teachers, students, and possibly parents; the changes in work practices that would occur for teachers; and potential for conflicting incentive systems that are likely to undermine the effort before it really could get underway. Despite the political argument that it is imperative to expand K–12 students' access to the Internet by allocating resources, policy analysts can use the findings of SI to bolster counter-arguments that there is a range of costly and possibly disruptive consequences that might accompany such a policy decision. Analysts would be able to clarify important and mostly tacit assumptions underlying the political argument about the abilities of school sites to keep their equipment "up and running"; the abilities of teachers to effectively integrate resources that are available on the Internet into a richer, inquiry-oriented curriculum; and the ways that K–12 students will use these resources when they are available.

4.2 A Historical View of Social Informatics-Oriented Policy Analysis

In this section, we examine historic and current ICT policy analysis organizations in the United States and Europe (the principal loci of ICT policy making during the past decade) to determine where SI thema have a presence, and to suggest places where they could be of benefit.

4.2.1 U.S. ICT Policy (1970–Present)

For the past thirty years some government organizations, especially in the United States and Western European countries, have been systematically publishing policy studies and policy-relevant studies about the impact of ICTs on economies and society (see Table 4.1, Some Major ICT Public Policy Analysis Organizations or Programs). In this section, we will focus upon those organizations and programs that have produced a significant number of ICT policy analyses or ICT policy relevant research.

4.2.1.1 U.S. Congress's Office of Technology Assessment

Studies conducted by researchers at the Office of Technology Assessment (OTA) in the United States between 1980 and 1995 underscored the need to understand the relationship between ICT developments and society. The OTA was created as a Congressional office in 1972 to provide research to Congress. The OTA's scope was very broad and included energy, transportation, international security, space, telecommunications, commerce, health, education, and the environment. It was designed to provide nonpartisan "objective" scientific and technical analyses for Congressional committees. The OTA was governed by a bipartisan Congressional technology advisory board that approved all of its substantial studies. The studies were conducted by a full-time professional staff that created stakeholder panels for each study and often hired consultants and university-based researchers to gather and analyze new empirical data. OTA became an important model for helping to inform politicians about the likely consequences of their legislation, and it stimulated the development of similar programs in a few Western European countries. Bimber (1996a) observes that:

> As is often true about the role of substantive expertise in the policy process, the effect of OTA's work was limited almost

Table 4.1 Some Major ICT Public Policy Analysis Organizations or Programs

Organization	Reports to	Period of Activity	Web Site
Office of Technology Assessment (OTA)	U.S. Congress	1972-1995	http://www.wws.princeton.edu:80/~ota/
Computer Science and Telecommunications Board (CSTB)	National Research Council — Commission on Physical Sciences, Mathematics, and Applications	1987-present	www7.nationalacademies.org/estb
President's Information Technology Advisory Committee (PITAC)	Executive Office of the President	1998-present	http://www.hpcc.go/pitac/
National Telecommunications and Information Administration (NTIA)	U.S. Department of Commerce	1978-present	http://www.ntia.doc.gov/
Secretariat for Electronic Commerce	U.S. Department of Commerce	1997-present	http://www.commerce.gov/egov.html
Programme on Information and Communication Technologies (PICT)	Economic and Social Research Council (ESRC) (U.K.)	1985-1995	http://www-rcf.usc.edu/~wdutton/pict.htm
Information Society Promotion Office (ISPO)	Information Society Activity Center of the European Commission	1994-present	http://europa.eu.int/information_society/index_en.htm

exclusively to early stages of policy-making, during agenda-setting processes. Staff and key committee members used OTA work in judging the salience of problems, in framing issues, and in identifying policy options—typically well before bills were even introduced. Legislators rarely drew the agency into the more publicly visible processes of debating bills, voting, and publicly explaining decisions. This fact contributed to OTA's low profile inside Congress and especially outside of it.

OTA produced approximately 50 book-length reports about ICT developments including studies of Federal government information systems (for policing, social security, tax administration, and postal services); ICTs in healthcare, finance, and education; privacy issues; workplace issues (from manufacturing shop floors to offices); and communications infrastructures in rural and metropolitan areas. SI researchers participated in some of these studies as panel members or as contract researchers. However, OTA's ICT panels that examined private sector developments often seemed to be dominated by various commercial interests. One report, *Critical Connections: Communication for the Future*, (U.S. Congress, Office of Technology Assessment, 1990) is worth noting for the ways in which it integrates the key SI thema listed in Figure 4.1 (p. 54). The summary of the report acknowledges dramatic changes in the U.S. communication system and views these changes as being potentially positive for society. However, it recognizes and articulates clearly that rapid technological change can also bring about negative consequences (U.S. Congress, Office of Technology Assessment, 1990, p. 3):

> New technologies hold promise for a greatly enhanced system that can meet the changing needs of an information-based society. At the same time, however, these technologies will undoubtedly generate a number of significant social problems. How these technologies evolve, as well as who will be affected positively or negatively, will depend on decisions now being made in both the public and private sectors.

In the discussion about policy issues and congressional strategies, *Critical Connections* states that the first priority for consideration is "equitable access to communication opportunities." Early OTA research presciently indicated that "changes in the U.S. communication infrastructure are likely to broaden the gap between those who can access communication services and use information strategically and those who cannot" (p. 11). In present discourse, this gap is often referred to as the "digital divide" and is assumed to be made up of "information haves and have-nots" (in Europe, "social exclusion").

Those on the wrong side of the divide include the urban and rural poor, those isolated from technology because of geographic location, and other groups of individuals who, because of economics, abilities, or location, cannot readily access ICTs. (Pew Internet Life Project, 2003a, 2003b, 2003c). Many technologists believe that reductions in computing costs will alleviate this gap. However, *Critical Connections* indicated that this gap would probably widen because of increasing costs associated with the development of new ICTs as well as communications policies that could affect how ICT developers and providers conduct business in the future. Current debates over the concept and nature of and possible solution to the digital divide seem to be playing out along lines articulated in OTA's work on ICTs in the 1990s (Cooper, 2002; Donnermeyer & Hollifield, 2003; Friedman, 2002).

In addition to making society rather than technology the focal point of ICT policy making, *Critical Connections* also provides a conceptual framework for analyzing policy issues affected by the rapid emergence of ICTs. This framework includes two opposing perspectives: one views technology as shaping society, and another views societies as shaping the technologies and technological configurations that they adopt (OTA, p. 34). Although "conceptual frameworks" are common in scholarly publications, their incorporation in this OTA report helps its audience understand the relationship between ICTs, policy, and society. Like research informed by SI, it also identifies the stakeholders involved and the positive and negative aspects of ICT impacts, and it foreshadows potential policy decisions.

4.2.1.2 The Computer Science and Telecommunications Board

Many academics are aware of the National Academy of Sciences (NAS), and view it as a highly prestigious scientific society, but fewer know that it was not founded to serve as an elite scientific society. It was established in 1863, during the U.S. Civil War, to help advise the U.S. Government on scientific and technical questions such as appropriate standards for weights and measures. Over the last 140 years the organization has expanded its structure (with the National Academies of Engineering and Medicine, the National Research Council (NRC) to organize its research, and the National Academies Press to publish its reports). The range of topics that the NRC examines now includes ICT innovation and its socio-technical aspects.

The NRC's Computer Science and Telecommunications Board (CSTB) is the primary organization within the National Academies that examines ICT issues. Over the last fifteen years, it has conducted several dozen systematic studies and issued a comparable number of book-length reports. Given its breadth, it is difficult to summarize the scope of the CSTB's charter in a few sentences. It is broader than was the OTA's scope because it includes mandates such as having to periodically assess the health of Computer Science and Computer Engineering as academic disciplines. Some recent reports examine the role of the U.S. Library of Congress in an era when many documents are published in digital formats, ways to improve the reliability of health care information that is available on the Internet, programs of social and technical research intended to advance our knowledge about how to make Internet access easier for "ordinary people," ways of adjudicating intellectual property rights in digital environments, and ways to improve the public accessibility to electronic government information (for a listing of available reports, see http://www7.nationalacademies.org/estb/publications.html).

With the demise of OTA, the CSTB is arguably the most significant ICT policy analysis organization in the United States. There are numerous important differences between the OTA's organization and processes and those of the CSTB. In brief, the OTA's studies were subject to Congressional approval, supported by Congressional funding, conducted primarily by a full-time professional staff with the assistance of occasional contract assistance, and guided by study panels who represented politically identified stakeholders. In contrast, the CSTB's studies are subject to approval by scientifically-oriented NRC review boards, funded by Federal agencies (i.e., Defense Advanced Research Projects Agency, Air Force Office of Scientific Research, National Institute of Standards and Technology, National Science Foundation, and the Office of Naval Research) and industrial firms (i.e., Cisco Systems, Sun Microsystems, Hewlett-Packard, and Time Warner), carried out on a voluntary basis by prominent scientific and technical experts, and guided by the CSTB. The CSTB's rotating membership is composed primarily of computer scientists and computer engineers from research universities, and a few industrial computer scientists who have made noted technical contributions to the discipline. Finally, OTA reports avoided making policy recommendations because policy making was the domain of their Congressional sponsors. In contrast, CSTB reports often advocate policy actions.

The CSTB staff argues that their reports change "the way people think about information technology and public policy" (Computer Science and Telecommunications Board, 2004). They take credit for advancing action recommendations that have often been adopted and adapted by the Federal government and some commercial firms. The CSTB also claims a unique and important role in altering the perceptions, beliefs, and actions of many academic computer scientists (Computer Science and Telecommunications Board, 2004):

> CSTB has additional impact as an institution: service on the Board and on the committees it oversees has expanded the public policy awareness and public service contribution of hundreds

of computer scientists, while enhancing the insight into information technology of the other experts it engages. Several of these individuals have adapted their own research programs and undertaken new and more substantial public service commitments as a result of the introduction and education they associate with CSTB. As public attention to the intersections of information technology and public policy grow, this human capital will become even more valuable.

CSTB's reports have been discussed in positive terms by senior Federal staff members who deal with ICT policy, program managers in research agencies (such as the NSF), and academic computer scientists. The reports are not as highly valued within the SI research community because they tend to emphasize a more dated view of ICT use in social contexts rather than the more current view of ICTs as socio-technical configurations and accomplishments. One popular CSTB study, *Information Technology in the Service Society* (CSTB, 1994), illustrates some of these limitations. This ambitious report examines the problem of measuring productivity changes that may result from the adoption and implementation of new information technologies. While U.S. service companies spent more than $750 billion in the 1980s just on computer and communications hardware, official productivity statistics had only grown an average of 0.7% a year. The study attempted to explain this paradox and report why standard measures of productivity are inadequate. The report includes a careful study of productivity statistics and some discussions of the economic changes resulting from computerization in organizations. The report seems to uncritically assume that computerization *must* improve productivity in organizations, yet the authors observe that "For the most part these productivity measures do not reflect aspects of service quality, such as speed or convenience that are affected by information technology, nor the alternative cost of what would have happened without it" (CSTB, 1994, p. 75).

However, several brief discussions within the report reflect the Social Informatics finding discussed earlier that the value of ICTs usually comes when social practices are changed along with the technologies; changing (or adding) technologies alone rarely produces many transformative benefits. The report cites an example provided by Raymond Caron, President of CIGNA Systems, that ICTs designed to save insurance underwriters work proved to be of little value because there was no effort to "design change in the overall process" of underwriting work (CSTB, 1994, p. 170). In one passage, the report picks up the Social Informatics finding and notes that:

Indeed, the full realization of benefits from using IT generally requires not just an extensive investment in hardware, but a complete overhaul of the firms' traditional organizations, systems practices and culture. ... A majority (60 percent to 80 percent) of companies interviewed by the committee found that the use of IT had an impact on their organizational structure (e.g., changing spans of control, facilitating organizational flattening, or encouraging use of self-directed teams). But very few made a full transition to supporting their new organizational structures with both new customer-oriented measures of performance and new reward systems. (CSTB, 1994, p. 180)

The report goes on to observe that:

Direct and intimate user involvement in the specification, design, and implementation of IT systems was a strong contributing condition for success in the vast majority (over 85 percent) of the companies interviewed. (CSTB, 1994, p. 181)

However, these ideas do not percolate into the report's concluding chapter or recommendations. Many readers could readily interpret this report as concluding that ICTs improve organizational productivity, but that the national statistics are misleading and new forms of measuring commercial performance need to be devised.

It is ironic that the NRC published another report about ICTs and productivity in the same year: *Organizational Linkages: Understanding the Productivity Paradox* (CBASSE, 1994). This report describes a study conducted by the NRC's Social Science Commission and foregrounds some of the SI insights that are mentioned, but buried, in the CSTB's report. Its authors try to break new ground by better understanding how organizational practices and coordinated change programs are integral to creating value from ICTs. This report does more to help make sense of the productivity paradox as represented in the survey data collected by the CSTB than does the CSTB's own report.

Some of the analytical limitations of CSTB's reports seem to derive from their tacit role in helping to legitimate the discipline of Computer Science within the NAS. Further, the CSTB is composed of elite computer scientists whose technical and scientific accomplishments help to anchor their appointments. Many of them have a deep concern that ICTs be developed and used in socially benign ways, but have little grounding in the SI research literature. They are predominantly people who would be viewed as broadminded by their academic and industrial colleagues. However, the question at hand is whether the CSTB's world-view is sufficiently engaged with the relevant SI research so that it can effectively lead the nation in how to view ICTs in the social world.

Further, the computer science research community, which influences the staffing of the CSTB and its studies, has generally avoided a close study of the actual uses of the kinds of ICTs that organizations and households routinely use. King (1997) makes this point in an appendix to another CSTB report about designing ICT interfaces for "ordinary people":

> Important aspects of research into group, organizational, and institutional usability have been under way for many years. Although largely ignored by the computer science research community, the vast range of economically vital computing applications in organizational information processing have drawn much attention from researchers in management information systems, library and information science, medical informatics, and other fields. Transaction-processing systems, which remain among the largest and most complex computerized information systems, were made possible only by careful study and learning-by-doing design to meet interface needs at the individual, work group, organizational, and institutional levels. To pick just one case in point, designers of the airline reservations system, which literally revolutionized air travel, had to overcome numerous complicated problems at all social levels, including being modified to comply with court-ordered remedies against unfair competitive practices. Similar stories can be told regarding credit data-reporting systems, financial accounting and reporting systems, personnel management systems, computer-integrated manufacturing systems, and so on. The lesson here is that a great deal of useful information on the development of effective interfaces at the higher social levels is available in the applications-oriented research communities.

As was indicated here, the CSTB plays a very important role in investigating ICT public policy in the U.S. today. It is arguably the most likely organization to bring relevant SI insights into national policy discussions of society and technology. Like all organizations, the CSTB is imperfect. Unfortunately, some of those imperfections significantly limit its ability to effectively draw upon the most relevant research.

4.2.1.3 President's Information Technology Advisory Committee

Today several executive branch organizations play significant roles in shaping the current technology development and ICT policy agendas in the U.S. Government: the Executive Office of the President (with its Office of Science and Technology Policy and various groups that report to it), and two offices in the U.S. Department of Commerce. One of the groups that advises the White House is President's Information Technology Advisory Committee (PITAC), whose membership includes twenty-six prominent academic computer scientists and ICT industry leaders. PITAC was charged with providing an independent assessment of the Federal government's role in information technology research and development. PITAC originally had a two-year tenure (which has since been extended until 2005) and issued a major report early in 1999. This report led the White House to propose a $336 million increase in Federal ICT research.

PITAC organized its inquiry through several subgroups. Their major foci and expertise lay in technologically focused topics, such as scalable ICT infrastructures and high speed computing. However, there was also a Socio-Economic and Workforce Impacts (SEW) panel. PITAC's SEW panel included a key SI argument in its overview, namely that ICTs have both enabling and constraining effects on social settings. The committee wrote, "we must understand the transformations and potential dislocations affected by technology adoption and diffusion" (PITAC, 1998). In an interim mid-1998 report, SEW illuminated the processes and consequences of ICT use and again underscored the need for more empirically anchored research:

> Much more social science research on the impact of IT on our society is needed to inform ongoing debates and policy decisions on IT-related issues. Such research can also help IT researchers develop technical solutions for some difficult policy problems (e.g., development of new metadata tagging standards and micropayment technologies for managing intellectual property). Moreover, insights derived from social science research may be able to contribute to the better designs of information systems. The design of GroupWare, for example, should be driven by research on how groups of people share information and make decisions. (PITAC, 1998)

Social science research differs from SI research. Social Informatics is not a stand-alone discipline, like Sociology or History; its thema cut across multi-disciplinary boundaries. Because the majority of the PITAC members have business, computing, or engineering backgrounds, it is unlikely that they are aware of SI research. However, their recommendations for possible research include many areas presently being investigated by social and organizational informaticians. These areas include, but are not limited to, transformations of social institutions by ICTs, sustainable use of large ICT infrastructures, electronic groups and communities, and barriers to ICT diffusion. For example, SEW focuses on the design of GroupWare and how groups of people share information and make decisions. For about fifteen years, social and organizational informaticians have researched the nature of group work and information sharing, as well as the use of ICTs that support teamwork (such as Lotus Notes).

4.2.1.4 U.S. Department of Commerce

In the U.S., the Clinton-Gore administration enthusiastically supported the widespread use of the Internet for numerous education, scientific, civic, and commercial applications from the time it took office in 1992. By 1997, the White House had articulated a framework for global electronic commerce (e-commerce) (Clinton & Gore, 1997) that advanced five major policy principles:

1. The private sector should lead (e-commerce development).

2. Governments should avoid undue restrictions on electronic commerce.

3. Where governmental involvement is needed, its aim should be to support and enforce a predictable, minimalist, consistent, and simple legal environment for commerce.

4. Governments should recognize the unique qualities of the Internet.

5. Electronic commerce over the Internet should be facilitated on a global basis.

U.S. Commerce Department reports about a '"digital economy" and e-commerce are much more advocacy documents than the kind of objective technology assessments that were supposed to be produced by Congress's OTA. These reports take principles, such as these five, as foundational. In 1998, the U.S. Department of Commerce published *The Emerging Digital Economy* (Margherio, Henry, Cooke, & Montes, 1998). It focuses principally on e-commerce and its economic impact on organizations, the transfer of goods and services, consumers, and ICT workers. The report places emphasis on expanding the Internet to ensure that enough bandwidth is available to both business and consumers so that both can take full advantage of e-commerce opportunities. It exhorts policy analysts to eschew implementing any Internet taxes, which would inhibit the free flow of e-commerce. The report also acknowledges that a growth in the digital economy will potentially bring about some downsides for society:

> The digital economy may bring potential invasions of privacy, easier access by children to pornographic and violent materials and hate speech, more sophisticated and far-reaching criminal activity and a host of other as-yet unknown problems. (p. 51)

However, the report is generally decoupled from the complexities that organizations face in effectively moving online. Most of the projects are described in terms of a series of tasks, and give us little clue about how organizations changed to accommodate new practices. Improvements in organizational subsystems are treated as organization-wide gains. For example, a description of the way that General Electric's Lighting Division developed an online procurement system focuses upon the efficiencies in the procurement department (faster orders, 30 percent cost reduction, and 60 percent staff reduction). However, there is no account of the costs of developing the online procurement system, deploying and maintaining numerous new workstations in the Lighting Division, training those who request materials ("the internal customers") to correctly specify orders online, to effectively use the online forms with digital drawing attachments, and so on. There may still be important net savings after these costs are factored in, but the cost reductions would not be so dramatic. The magnitude and characteristics of the co-requisite organizational changes would also be clearer.

Most seriously, this expanded view suggests that IT should not be conceptualized simply as a "tool" which can be readily applied for specific purposes. GE Lighting's online procurement system has important features as a complex technological system in which the orchestration of digitized product drawings and purchase orders has to be synchronized. It has important social properties regarding the authorizations to initiate an electronic purchase order, the control over product drawing versions that have been subject to engineering changes or manufacturing changes, and so on. In short, organizational researchers have found that systems like this are better conceptualized as "socio-technical networks" than as tools. In practice, the boundaries between what is social and what is technological blurs because some of the system design encodes assumptions about the social organization of the firm, in this case GE Lighting.

A different kind of example that could have enriched this report comes from the experience of Charles Schwab and Co. when they developed an online trading operation (eSchwab) in 1995–1996 (Schonfeld, 1998). Like many firms, Schwab initially set up a new small division to develop the software, systems, and policies for eSchwab. To compete with other Internet brokerages, Schwab dropped its commissions to a flat fee that was about one-third of its average commission. Schwab's regular phone representatives and branch officers could not help eSchwab customers. eSchwab customers were allowed one free phone call a month; all other questions had to be e-mailed to eSchwab. While over a million people rapidly flocked to eSchwab, many of these customers found the different policies and practices to be frustrating. In 1997, Schwab's upper managers began integrating eSchwab and "regular Schwab." This integration required new, more coherent policies and training all of Schwab's representatives to understand e-trades. It also required the physical integration of eSchwab's staff with their "jeans and sneakers" culture into the offices of regular Schwab staff with a button-down "jacket and tie" culture. One result was the development of a more flexible dress code in Schwab's headquarters.

eSchwab has been discussed in some business articles as a tool or a technological system, but the policies and procedures for any trading system, including pricing, trade confirmations and reversals, and advice, are integral to its operation. These are social practices without which there is no eSchwab. Consequently, the standard "tool view" is insufficient for adequately understanding the design of eSchwab, its operations, and, consequently, the character of the organizational change required to develop this line of business (Kling & Lamb, 2000).

The Department of Commerce's National Telecommunications and Information Administration has issued three reports about the social distribution of telephone, computer, and Internet access in the United States (NTIA, 1998, 1999). The studies are based on data collected by the Census Bureau from 48,000 U.S. households and seriously engage the data. According to the 1999 report:

- Between 1997 and 1998, the divide between those at the highest and the lowest education levels increased 25 percent and the divide between the highest and the lowest income levels grew 29 percent.

- Households with incomes of $75,000 or higher are more than twenty times as likely to have access to the Internet than those at the lowest income levels and more than nine times as likely to have a computer at home.

- Whites are more likely to have access to the Internet from home than Blacks or Hispanics have from any location.

- Black and Hispanic households are approximately one-third as likely to have home Internet access as households of

Asian/Pacific Islander descent, and roughly two-fifths as likely as White households.

- Rural areas are less likely to be connected than urban areas. Regardless of income level, those living in rural areas are lagging behind in computer and Internet access. At some income levels, those in urban areas are 50 percent more likely to have Internet access than those earning the same income in rural areas.

Since ICT costs were generally declining between 1997 and 2003, the expected decrease in the costs of acquiring PCs and Internet service cannot be expected to automatically redress these increasing differences. It is an open question about how these differences—the digital divide—translate into other inequalities, such as access to better jobs, business starts, improved education, and other social goods. Even so, these NTIA reports provide a sound empirical basis for informing ICT policy.

However, the policy analysis arm of the U.S. Department of Commerce that has developed the Digital Economy reports lacks the kind of SI research capacity that OTA demonstrated in its best studies. Nor does it know how to articulate the kind of empirically grounded, constructive skepticism that added the question mark to the Virtual Society? Programme.

4.2.2 Private ICT Research Institutes in the 1990s

Over the past decade, a number of private policy-oriented organizations have systematically examined ICT-related issues. These include the Internet Policy Institute, the Benton Foundation, the Electronic Privacy Information Clearinghouse, and the Progressive Policy Institute. At the same time, many other organizations have articulated and tried to publicize ICT policy preferences for topics intersecting with their areas of concern. For example, the American Library Association has developed policy positions and given testimony to the U.S. Congress on topics such as copyright, using software filters to restrict children's access to Internet sites, and subsidies for libraries in disadvantaged areas to access the Internet (see http://www.ala.org/oitp).

The Progressive Policy Institute's (PPI) *The Internet and Society: Universal Access, Not Universal Service* is concerned with the impact of the Internet on society. The authors see universal Internet access (as opposed to universal service) as the goal policy analysts should strive for because universal access ensures that people who want or need to access the Internet can do so. They suggest that the goal of universal access could be met if the government provided Internet access in places such as libraries, rural health centers, and schools, all of which qualify for the U.S. government's E-Rate Program.

Moving toward a universal service model is not recommended, because the authors do not find enough evidence to indicate that widespread demand for Internet access exists at present. They also present evidence to support the claim that "the real technology divide in the nation is one caused by technology-related changes at work and increases in income inequality, not differences in ownership of various technology devices" (Progressive Policy Institute, 1998, p. 7).

These reports from the U.S. Department of Commerce and from the PPI are anchored in the concept of the "New Economy." According to the PPI, the "New Economy" has been emerging in the United States for the past fifteen years and is shaped by information technologies and new communications networks like the Internet (Progressive Policy Institute, 1998). This conceptualization embraces technological determinism by seeing the "New Economy" as being shaped by ICTs and, apparently, nothing else. Supporters of the "New Economy" also appear unwilling to engage in discussion about broader social concerns such as the impact ICT deployment will have in the workplace, in the home, and on the individual.

4.2.3 European ICT Policy Analysis (1985–Present)

ICT policy analyses reflect the diversity of European national governments. A few countries, such as the Netherlands, have developed national technology assessment programs that have been influenced by the OTA. In contrast, many European countries produced national-level "information society" proposals in the mid-1990s, even if they had little other systematic ICT policy analysis capacity or effort. For simplicity, we will focus on one major national research program and ICT policy analysis in the European Commission.

4.2.3.1 The United Kingdom's Programme on Information and Communication Technologies

The Programme on Information and Communication Technologies (PICT), a ten-year (1985–1995) SI research initiative based in six university-based centers in the United Kingdom, loosely parallels some of the research that is reported in OTA's ICT studies. Unlike OTA, whose studies were driven by the agendas and foci of a national political body, PICT was a sustained social science research initiative. PICT researchers examined and published numerous books and hundreds of articles about the social shaping of ICTs, as well as examining associated policy issues (Dutton, 1996, 1999; Kubicek, Dutton, & Williams, 1997). Dutton, PICT's last director, characterizes the main themes of PICT research as falling into four major areas: 'Production' (the social shaping of technology), 'Utilization' (organizations, management, and work), 'Consumption' (reactions of the individual consumer and citizen), and 'Governance' (policy and regulation). Much of the research conducted under the first three rubrics was policy-relevant research. Although PICT's research should inform ICT policy making, we have noted that these influences take many years to develop. Even the U.S. Congress's OTA, which was much more closely coupled to a policy-making body, was not well integrated into the full stream of Congressional policy analysis and policy making.

In an integrative review of PICTs research, Dutton (1999) discusses his search for an all-embracing integrating theme underlying the concept of the social "shaping of tele-access." His book illustrates how this concept:

> ... can help integrate the findings of research on ICTs across the social sciences generally. Many social and economic issues—ranging from issues of information inequality, privacy, and censorship, to the role of the Internet, and information superhighways in economic development—can be better understood if viewed as products of a process that is quite literally reshaping social and economic access in this digital age of new ICTs. (Dutton, 1999, p. 27)

The analyses that underlay the social shaping of tele-access are basically similar to those discussed in Chapter 2, which examined how ICTs should be viewed as configurable socio-technical networks. The analysis in that section drew upon some of the PICT research, as well as research conducted in the U.S. PICT research has been able to be more fundamental and cumulative than OTA's, which was driven by the shorter-term interests of members of Congress and their staffs, and also influenced by practitioners who served on study panels. In contrast, PICT was much more influenced by the values, norms, and processes of the British social science research communities.

PICT disbanded in 1995 and was superceded by a substantial new "Virtual Society?" research program that addresses three major policy-relevant themes (Virtual Society?, 2000):

1. Skills and Performance: the impact of new electronic and communications technologies on human and organizational potential, performance and learning

2. Social Cohesion: the role of new electronic techniques in relations between people and in modifying processes and degrees of social inclusion and exclusion

3. Social Contexts of New Electronic Technologies: the
    changing social contexts and factors influencing the
    transformation and adoption of electronic technologies

Both PICT and the Virtual Society? research programs have had an interesting critical edge to them. The Virtual Society? (2000) description claims
that it "benefits from research which retains some (analytic) scepticism
about the claims made for the new technologies. The "?" in "The Virtual
Society?" signals this analytic stance." Unfortunately, there is no program of
policy-relevant research of comparable scale and quality funded in North
America.

4.2.3.2 European Commission's Information Society Project Office in the 1990s

The Information Society Project Office (ISPO) is the European
Commission's key organization for developing studies of ICT policy. Like the
OTA, it is part of an official policy-making organization. Unlike the OTA, it
does not have a substantial staff that is grounded in Social Informatics.
Instead, it relies upon the work of blue-ribbon committees and volunteers to
produce many of its reports. The ISPO was originally named the Information
Society Project Office when it was founded around 1993 and changed its
name to the Information Society Promotion Office in November 1998.

It is worth noting that the term "information society" as used by the ISPO
tends to emphasize the development of new ICTs within market arrangements. There is a well-developed literature about the concept of an information society that offers a much richer conception (Bell, 1973, 1980; Webster
1995) that is not incorporated into ISPO's discourse. One of the earliest theorists of an information society, Daniel Bell (1980), identifies three elements of
a post-industrial/information society: (1) the change from a goods-producing
to a service society; (2) the centrality of the codification of theoretical knowledge as a driving force in society; and (3) the creation of "intellectual technology" (such as management science) as key elements of production. None
of these criteria could be satisfied primarily by the development of new ICTs.
The shift to a service society is a shift in the industrial mix in an economy; the
codification of theoretical knowledge and the creation of new intellectual
technologies is labor and skill intensive. Bell's argument is that knowledge is
the key strategic resource in information societies and replaces "labor as the
source of added value in the national product" (Bell, 1973, p. 506). In substi-
tuting this "knowledge theory of value" for Marx's "labor theory of value,"
Bell's information society represents a major break with the industrial era.
Bell's theories of an information society have stimulated considerable
debate that goes well beyond the scope of this book. The key point is that the

working concept of an "information society" that is reflected in most of the ISPO reports is much more technologically focused than is the concept's scholarly version.

One early and widely circulated report, *Europe and the Global Information Society*, was written in 1994 by a committee chaired by former European Commission Vice-President, Martin Bangemann. The report has a heady optimism and excitement that also characterized some early Clinton-Gore reports about a National Information Infrastructure in the U.S and emphasizes the importance of markets in fueling an information revolution that can benefit Europeans. The Bangemann report proposed large public investments in ten application areas, including teleworking, distance learning, networks for universities and research centres, telematic services for small and medium enterprises, road traffic management, an integrated European air traffic control system, health care networks, electronic banking, a trans-European public administration network, and city information highways. Like the U.S. Department of Commerce's 1999 report, *The Emerging Digital Economy*, it assumes that these kinds of applications can be readily built and produce tremendous social value if there is simply sufficient funding.

The Bangemann Report was an important and easily readable polemic to help stimulate discussions of market alternatives and trans-European ICT developments. However, its breathless quality allows no space in which to ask about the complexities of activities, such as distance learning. Its authors ignored the best empirical research about ICT developments and their polemic stance hardly allows the "question mark" that is integral to the U.K.'s Virtual Society? Programme. The Bangemann Report motivated some within European ICT policy-making circles to call for policy that is more socially embedded. The ISPO formed a High-Level Expert Group (HLEG) that encouraged future researchers and policy analysts to move away from a technologically deterministic focus toward one that embraces concepts of social embeddedness.

HLEG's final policy report, *Building the European Information Society for Us All* (April 1997), identifies the problem of policy discourse within the EC regarding ICT implementation, namely that it had "been dominated by issues relating to the technological and infrastructure challenges." However, continued work that looked closely at how these "challenges" would impact society led HLEG to conclude "the field has expanded rapidly, with social aspects of the emerging IS [information society] moving to the top of the policy agenda." HLEG builds on this by creating an outline or "vision" for future policy discourse and by providing recommendations for implementation.

Like the OTA's report, *Critical Connections* (U.S. Congress, Office of Technology Assessment, 1990), the HLEG report views the rapid emergence of ICTs as both a potential blessing and a potential curse. It describes the need to find a policy model that "avoids social exclusion and creates new opportunities for the disadvantaged" (HLEG, 1997). It also makes an important distinction between data, information, and knowledge: "The generation of unstructured data does not automatically lead to the creation of information, nor can all information be equated with knowledge." This advances the notion that the emergence of ICTs does not necessarily make for a "wiser" society. Further, HLEG (1997) overtly rejects the framing of any policy regarding ICTs in the language of technological determinism:

> The social integrationist vision that the HLEG espouses explicitly rejects the notion of technology as an exogenous variable to which society and individuals, whether at work or in the home, must adapt. Instead, it puts the emphasis on technology as a social process.

With regard to future ICT policy development, HLEG states that "the new ICTs embody a radically different set of parameters for potential growth and development opportunities." It identifies ten policy challenges that provide a "broad agenda for policy action involving a range of actors." Some of these challenges include social exclusion, knowledge and skill acquisition, and bridging geographical distance. The report concludes on an optimistic note by identifying new social possibilities that ICTs enable but does comment that:

> New structures are nevertheless needed which reflect the new possibilities of the IS (information society) and which permit the development of demand for new ICT-based services. Organisations and structures of the past are not necessarily going to meet this demand. We could be defensive about this and try to hang onto the treasured aspects of the old systems, or we could try to define a better more open path, which overcomes some of the centralisation and authoritarian aspects of the traditional bureaucracies and governance structures to be found in Europe. (HLEG, 1997)

Unfortunately, the HLEG report has been removed from the ISPO's Web site. Its complex social vision probably is a bit discordant with the "can do" tone of other ISPO reports. In December 1999, the European Commission launched an initiative entitled "eEurope—An Information Society for All," which proposed ambitious targets to bring the benefits of the Information Society to all Europeans. The initiative focuses rapidly accelerating broad public ICT access in ten priority areas, from education to transport and from health care to the disabled. The report sets aspiration levels for the next three years. Unfortunately, the report doesn't have any serious or systematic empirical reference points, such as we find about the digital divide in the NTIA's report series. In short, the ISPO seems to be a promotion office without a significant research enterprise that is coupled to its proposals and helps to shape them.

4.3 ICT Policy Analysis in the Next Decades

What emerges clearly after studying current and historic ICT policy is the infrequent connection of current ICT policy discourse to research and theory informed by Social Informatics. This situation, of a disconnect between research and policy development, is very common. In the case of ICT developments, the research often finds that developments are more difficult to utilize than many participants expect, and that workable ICTs often involve important changes in social practices. This makes ICT-focused social plans such as those promoted by the ISPO to be problematic. In contrast, the now disestablished OTA's research capabilities, as imperfect as they were, were arguably the most systematic and institutionalized efforts to bring SI knowledge to bear on a variety of ICT policy issues.

It is conceivable that the policy discussions, such as those about ICT access, will focus primarily on technological and economic practices rather than on social ones. This would be regrettable because governments have placed so much emphasis on access being the lynchpin for providing more opportunities for all. Fortunately, this situation can be remedied by including an SI perspective in the discourse.

The examination of historic ICT policy analysis in the United States and Western Europe reveals that in the past, policy analysts (particularly those affiliated with OTA and PICT) strove to include concepts of social embeddedness as a primary concern in ICT policy making. However, current discourse seems to show greater interest in ICT development and its role in fueling the "New Economy." There are two possible explanations for this. The first lies in changes in the composition of research or committee teams who formulate ICT policy. OTA and PICT, for example, included many social science researchers on their respective teams. Consequently, much of their work looks closely at the impacts of ICTs on society and is framed in the language of social embeddedness. NCO and ISPO have a few of these individuals, but the majority comes from computer science, engineering, economics, and private industry. Although these people have deep domain knowledge within their respective fields, many may not be grounded in social theory. Therefore, it is not surprising that policy analysis and recommendations emanating from NCO and ISPO focuses more on technology development and economics than on social impact. Another explanation may have more to do with timing. Both OTA and PICT were organized before the Internet became such a worldwide phenomenon. Both organizations examined the relationship between technology (in a broad sense) and society.

ICT development has been transformed in the last decade. For example, the Internet is no longer solely the domain of universities and the government. With the advent of commercial ISPs (e.g., AOL and Earthlink) and graphical user interfaces (e.g., Netscape and Internet Explorer), the public has been able to access vast sources of information.

It is clear that any future discussion of ICT policy must include an integrated social focus, if publicly stated government goals of providing more opportunities and a better quality of life for society (locally and globally) are to be achieved. Building upon existing or commissioning new research informed by Social Informatics would help ICT policy analysts to ensure that new ICTs are shaped so that they benefit those who use them either at work or at home. It will also help policy analysts to better address questions such as: Does everyone need Internet access? Will the "New Economy" bring prosperity to all and eliminate the gap between information "haves" and "have-nots"? How will we resolve the IT worker shortfall? Will ICTs displace workers and, if so, what stopgap measures will be in place? Will ICTs give us a better quality of life? Answers to these questions cannot be provided simply, nor solely through SI research, but such research is an important contribution for developing sound answers.

CHAPTER 5

Teaching Key Ideas of Social Informatics

5.1 Why Teach Social Informatics?

In this chapter, we explain why ICT-oriented students should learn key concepts of social and organizational informatics. For this discussion, the ICT-oriented disciplines are those that educate students in the design, development, implementation, and support of ICTs, primarily in computer science and information systems programs.[1] This attention to the roles of social and organizational contexts in ICT-oriented curricula reflects the ongoing efforts in science, technology, engineering, and mathematics (STE&M) disciplines to develop students' abilities to use their technical educations more effectively (Freeman and Aspray, 1999; NSF, 1996, 2002).

This chapter is written for three audiences involved in ICT-oriented education: program administrators, curriculum committee members, and relevant educators. For all three audiences, we highlight the core concepts of Social Informatics and how these concepts add value to an ICT-oriented education (at both undergraduate and graduate levels). For program administrators, the chapter provides a comprehensive introduction to the importance and contemporary treatment of Social Informatics teaching. For curriculum committee members, this chapter discusses how and where Social Informatics concepts should be included in curricula. For educators, we discuss some good ways to teach Social Informatics theories and concepts in ICT-oriented disciplines.

The result of differences in the goals across the different types of ICT-oriented degree programs means that, while the Social Informatics principles are common, the examples and materials may differ by program. For example, information systems students typically learn some organizational informatics concepts as part of their course in systems analysis. Computer science students are often able to take elective computing ethics courses, in which they are presented with some Social Informatics material, but they usually have less access to organizational informatics ideas in their curricula.

More broadly, the professional computing community generally agrees that computing oriented students should learn Social Informatics concepts and theories. However, these Social Informatics concepts and analytic techniques are rarely taught in ICT-oriented curricula. There are several reasons why this is so, including concerns about computing curricula being the proper academic location for Social Informatics concepts, ignorance of Social Informatics literature and findings, and even latent hostility regarding certain findings about the negative effects of ICTs in some settings. The paradox between the value of Social Informatics and its provision in ICT-oriented curricula helps to frame this chapter.

This issue is discussed in this chapter as follows. First, we lay out the need for teaching Social Informatics. Second, we outline the basis for and objectives of a Social Informatics component in an ICT-oriented education. Third, we summarize the current teaching of Social Informatics and controversies with *teaching social aspects of computing*, and fourth, we provide some suggestions for how to present key concepts. Last, we provide both a summary of this section and some recommendations regarding constructing Social Informatics-informed curricula.

5.1.1 Social Informatics Teaching in the Context of Broad Trends in Science-Oriented Education

A technically/scientifically educated student is highly valued by contemporary organizations—and by society—yet the numbers of college graduates in most STE&M disciplines is declining (NSF, 2002). The National Science Foundation (NSF) data indicate that STE&M enrollments account for about 5 percent of the college-bound students in the United States, and that the total number of graduates is not growing but seems to be decreasing (particularly post the 'dot-com bust'). This decline is even more alarming when looking at the enrollments of females relative to males (Cahoon, 2003).

Over this same period of declines in computer science enrollments and little change in the number of information systems students, there has been a growing demand for computer-oriented students, especially for those who combine technical expertise with excellent organizational/interpersonal skills and knowledge. For example, the Information Technology Association of America's (ITAA) 2001 report suggested a shortage of 900,000 IT workers (though this projection was developed during the tail-end of the early 21st century dot-com boom). As noted in a more recent report commissioned by the U.S. Department of Labor, the Rand Corporation suggested that the number of information systems and telecommunications analysts positions will continue to expand in the foreseeable future as more firms and industries both engage in and expand their uses of IT (Karoly & Panis, 2004).

The dichotomous trend of a rising demand for ICT skills in the workforce and a simultaneously decreasing number of STE&M–trained college graduates (specifically ICT–oriented graduates) was first noticed in the early 1990s (NSF, 1996). This contradiction helped bring about a sense of a crisis regarding the value and basis of STE&M education (Abraham & Hoagland, 1995).

For example, over the past ten years, the NSF's assessments of teaching in STE&M disciplines highlight problems with instruction, relevance of material, and other curricular and pedagogical concerns (NSF, 1996, 2002). These reports encourage STE&M educators/curriculum to focus on:

1. The process of developing scientific knowledge through exploration, invention, and expansion

2. Developing and communicating scientific understanding through collaborative learning

3. Developing reflective judgment and the ability to apply creative, flexible, and analytic thinking to help solve problems

4. Understanding science/technology issues as they exist within social and cultural contexts (Carr, 1997)

Empirical studies of the time noted that most computer science graduates lacked any rigorous and empirically based exposure to Social Informatics concepts, findings, and analytic techniques (Huff & Martin, 1995). Since these reports were published, it appears little has changed in terms of the curricula, but awareness is growing that change must occur (Kling, 2003). More pointedly, the focus has turned recently to improving STE&M education in two fundamental ways: 1) improving the quality of instruction and engage students in STE&M-relevant material, and 2) making STE&M education more visibly relevant to the lives, work, and future careers of students.

Building on the evidence and guidance from the NSF and the recommendations included in the 1998 report on information technology in the United States relative to making scientific progress in ICT more directly relevant to society (PITAC, 1998), we see at least four reasons to include Social Informatics concepts and relevant analytic techniques in ICT-oriented student's' formal educational programs:

1. To provide a set of conceptual frameworks for computing-oriented students to organize and assimilate the social and organizational forces affecting the functionality embedded into ICTs

2. To help these students understand that the design, configuration, and implementation of ICTs is a socio-technical process—one that is charged with both social and organizational implications

3. To help these students develop a set of analytic techniques to help identify and evaluate the social consequences of ICT-based systems

4. To assist technically trained students in developing a more critical (by which we mean reflective, inquiry-focused, and problem-based) appreciation of the benefits and limitations of ICTs.

Fred Brooks, in his speech accepting an award for his pioneering software engineering work, observed that, "Hitching our research to someone else's driving problems, and solving them on the owner's terms, leads us to richer computer science research" (Brooks, 1996, pp. 64–65). Although the focus of this quotation is research, the implication is clear: ICTs are most valued when they are set into the social or organizational contexts in which the problem arose. ICT-oriented students learning about the potential roles that computer-based systems can play in organizations and society should be able to understand that (1) the design, development, and implementation of ICTs is not value neutral, and (2) that outcomes of using ICTs are often paradoxical. Students presented with a mechanistic and technologically deterministic perspective are often uncomfortable with the challenges to this simplified view that both personal experience and broad-scale evidence provide. An example of the power of pro-technology enculturation (and the potential to reify the value of ICTs) is reflected in Microsoft founder Bill Gates's (1995) comment: "Because the most efficient businesses have an advantage over their competitors, companies have an incentive to embrace technologies that make them more productive" (p. 135). This oversimplification confuses efficiency with embracing technology and does not address the challenges of linking uses of ICTs to productivity changes.

On a more practical level, ICT-oriented students need to learn Social Informatics analysis techniques in order to increase the likelihood that ICT-based systems will be used and be valued. Their absence leads to an increased risk of designing, selecting, or configuring systems that occasionally work well for people, occasionally are valuable, are sometimes abandoned, are sometimes unusable, and thus incur extra costs and inspire misplaced hopes (Kling, 2003).

If nothing else, the continued lost potential that arises from poorly designed, built, or implemented ICTs justifies the added effort required to teach Social Informatics concepts. The social and organizational influences on the processes of designing, constructing, and embedding ICTs into social/organizational systems are both subtle and far-reaching. ICTs are becoming more pervasive and more important to our work and personal lives; and at the same time, the subtle influences of design-oriented decision making regarding the nature of people, their social actions, and the institutional forces embedded into ICTs are also not clear-cut, or even easily accommodated. Social Informatics principles are becoming more important, not less, in ICT-oriented education. For example, we have known for years that the rising use of electronic work teams supported by e-mail and other computer-mediated communication systems highlights the difficulty that dispersed work groups have in reaching consensus, including the tendency to see less pro-social behavior (flaming) (Sproull & Kiesler, 1991). Yet, the ongoing uptake of ICTs by children and teenagers is being designed and deployed by people who have never been exposed to this science or the guidelines regarding pro-social design that arise from the findings of this work (Large, 2004).

ICT-oriented students should also be aware that human behavior (especially behavior in formal organizations) is paradoxically bounded. A consistent finding of Social Informatics research is that human behavior is often more understandable than would at first be perceived by a technologist (Lamb & Kling, 2003). Conversely, this behavior is also never as predictable as the direct effects models of ICT design/use suggest (Ackerman, 2000). Thus, designing ICT-based systems requires a depth of social and organizational understanding that demands practice and guidance to do so effectively. For more than a decade, we have known that simplified caricatures of organizational life relative to systems design, the type often found in introductory MIS texts, underrepresent the difficult and nuance-laden process of ICT design/implementation (i.e., Beath & Orlikowski, 1994). However, a reading of the emerging encoding of the software engineering body of knowledge (SWEBOK) suggests that this set of findings does not appear to have entered, in any significant way, the formal educational coursework in software engineering (Abran & Moore, 2004).

At a broader scale, discussions of Social Informatics and ethical issues allow computing-oriented students to engage in systematic analysis about effects and implications of ICTs at multiple levels. Methods to engage the socio-technical development of ICTs more directly have and continue to be an active area of scholarship (Avison & Wood-Harper, 2003; Checkland & Scholes, 1990; Huff & Martin, 1995). These techniques provide students with the means to both frame and explore Social Informatics and ethical topics in the design, development, and implementation of ICT-based systems. And although ICT-oriented accreditation bodies (such as the Accreditation Board for Engineering and Technology (ABET)[2] have begun to require that such topics be taught to computer science students, this active scholarship remains too separated from most computing-oriented curricula.

These four comprehensive purposes undergird the importance of presenting Social Informatics concepts and methods to ICT-oriented students, yet this population is heterogeneous and the various groups' needs differ. Three primary differentiators are:

- Disciplinary affiliations (as discussed earlier)

- The type of work these students will obtain upon graduation

- The level of education (e.g., bachelor's versus master's degrees)

An example of work differentiation is that information systems students will tend to find work focused on the early design and/or implementation and support of ICTs. Computer science students will tend to find work involving the design and constructions of ICTs—often in software houses or hardware manufacturers (and, thus, away from the life worlds of the people who will use them). ICT-oriented students from other disciplines may also differ by the locus of their work. For example, information science students may be involved in developing and/or managing specific types of ICTs. As the role of the Web in society expands, the development of Web-based systems and the emerging work category of "content provider" will further expand the set of disciplines that are "ICT-oriented." Like many of the growing number of ICT-oriented jobs, this work is tied to the organizational and social contexts in which these systems exist.

The differences in student's educational levels will also be reflected in their likely work after graduation. Both undergraduates and graduate students may work in the same area, but with differing levels of responsibility. For example, the information systems graduate student is often expected to take on a managerial or leadership role, whereas a computer science graduate student is often asked to be a technical leader or focus on specific technical issues.

5.2 Summarizing the Teaching of Social Informatics

In this section, we summarize current status regarding the teaching of Social Informatics concepts. In the first subsection, we discuss the intellectual roots of Social Informatics and a review of the current offerings about scope and content. In the second subsection, we discuss some issues with the current status of Social Informatics teaching with respect to current pedagogical guidance from both the computer science and information systems fields.

5.2.1 Current Status of Teaching Social Informatics

As we indicated at the outset, Social Informatics concepts and analytic techniques are important for ICT-oriented students. Currently, many universities offer a broad variety and number of rigorous Social Informatics courses at both the undergraduate and graduate levels in a variety of disciplines. For example, a simple search of the Web returns almost 1,000 pointers to syllabi (from courses offered in 2003 or 2004) that explicitly include Social Informatics content. These syllabi present a rich, diverse, and extensive sampling of potential topics, pedagogical approaches, projects, and other activities that are being used in Social Informatics-focused courses. Departments that offer Social Informatics courses include, for example, computer science, information systems, information science, and various social science departments such as sociology and communications. Typically, Social Informatics offerings in computer science departments are upper-division electives and often focus on "computers and society." These courses have been offered in many computer science departments since the mid 1960s (see Ashenhurst, 1972) and supporting textbooks are available (i.e., Agre & Schuler, 1997; George, 2003; Kling, 1996d). Further, as we noted, accreditation boards (such as ABET) are demanding this material be a part of formal educational programs.

The Social Informatics offerings in information systems, where the department is often housed in a business school, are focused primarily on organizational informatics issues. Most introductory MIS textbooks provide some coverage of organizational informatics (and increasingly touch on Social Informatics issues and/or ethics). Most systems analysis and design texts explicitly discuss organizational informatics issues in the context systems design. Further, the texts on change management and implementation of ICTs typically focus on organizational informatics issues. And courses in both human–computer interaction (HCI) and computer-supported cooperative work (CSCW) often include organizational informatics material.[3]

In the ongoing curricular efforts of the professional societies involved in business-school/information systems education, there is also explicit discussion about the need for (undergraduate) students to be educated on organizational informatics (and to a lesser extent, Social Informatics) issues: "Creating systems in organizations includes issues of innovation, quality, human–machine systems, human–machine interfaces, socio-technical design and change management" (Davis, Gorgone, Couger, Feinstein, & Longenecker, 1997, p. 7). Three (of the 10 recommended) courses (including two of the first three) have, as part of the course's goals, organizational informatics (and some Social Informatics) concepts and techniques.

Information science schools also teach some Social Informatics topics (mostly at the graduate level). Along with those courses centered on systems design (such as HCI and CSCW) are those that focus on information policy and engage issues such as access, copyright, and intellectual property.[4] This broad, and diverse, collection of course offerings has also led to a number of textbooks and anthologies that focus on Social Informatics issues in design, implementation, and use of ICTs. Within these three communities, we estimate that, as of late 2004, there are more than 300 distinct courses taught annually in North America that include some Social Informatics concepts.

Social Informatics concepts and literature are more integrated into the curricula of European (and Australian) computer science, information systems, and information science programs. This is particularly the case in Scandinavia, where participatory design is often mandated in union contracts. This geographic distribution also provides a source for rich cross-cultural comparisons of approaches to the design and use of ICTs (e.g., Braa & Monteiro, 1996).

5.2.2 Issues with the Current Status of Teaching Social Informatics

The general status regarding Social Informatics teaching seems both impressive and improving, yet the picture is much less benign when we examine what fraction of ICT-oriented students take these courses. There is no national-level empirical data regarding enrollments in such courses. This lack of empirical insight is further complicated by the heterogeneity of ICT-oriented students' educations, career paths, and work roles/locations.

A broad, simple, and disturbing summary is that, in most ICT-oriented programs, Social Informatics topics are rarely covered in any depth. When they are covered, Social Informatics concepts and analytic techniques are either compartmented in an elective course or simplified to such a level that they have little discernable value (or connection to the rigorous empirical research from which they came). So, the question becomes: If Social Informatics concepts are acknowledged as important, why are they not taught in most ICT-oriented programs?

For example, in most computer science programs there is no curricular requirement to have any Social Informatics courses. When they are offered, specific Social Informatics courses are usually an upper division (or, for graduate students, a higher level) elective. Often these courses are offered on an infrequent basis. In some departments, the value of these courses is degraded through the process of assigning the teaching of these courses to instructors or adjuncts who may not be versed in the rigorous empirical Social Informatics literature. Many Social Informatics courses are taught by committed faculty who are heavily invested in both the content of the course and overall importance of this literature in the education of their students.

When this is not the case, both the value of the topics and the centrality of their value to ICT-oriented students are undermined.

Although ABET recognizes the importance of Social Informatics concepts to computer science education, accreditation is not yet widespread. Further, the responses of those few computer science departments at Carnegie-rated Research Extensive and Research Intensive U.S. universities who have sought ABET accreditation provide some insight regarding how the Social Informatics component of a computer science education can be handled. At MIT, the computer science department employs external experts from other departments and schools in the area to staff its Social Informatics-related courses. The computer science department at Georgia Tech has chosen to develop (and maintain) internal expertise, having faculty who are invested in Social Informatics issues teach these courses. However, these topics do not appear to be taught in computer science at some universities (such as UC-Berkeley) although certification demands it. At UC-Berkeley, for example, an upper-division elective course is maintained in the catalog, but it has not been offered since the sponsoring faculty member retired and because the current faculty show no interest in developing (or contracting for) Social Informatics expertise.

Conversely, for students in information systems programs, Social Informatics concepts (albeit primarily organizational informatics) and analytic techniques are important aspects of their education. Thus, Social Informatics concepts are often taught in a number of common/core courses (such as the introductory course, the systems analysis course, and the project and/or change management course). Further, Social Informatics-related issues of ICT design, implementation, and use—and the resulting organizational and work-related changes—are currently extremely viable and fruitful areas for information systems faculty's research. This also encourages faculty to engage their students with Social Informatics concepts. Social Informatics content is also likely to be found in other departments in business schools (such as organizational behavior or economics). Thus, most information systems students are at least exposed to organizational informatics (and to some extent Social Informatics) issues in a number of classes.

Despite the general growth of Social Informatics courses as elective choices in many curricula, and the particular emphasis on organizational informatics issues in most information systems curricula, there are very few computer-oriented programs that require Social Informatics courses. Further, the example of how computer science departments in Carnegie-rated Research Extensive and Research Intensive universities deal with the Social Informatics and ethical analysis topics mandated by the ABET highlights three forces that affect the otherwise rosy picture regarding the teaching of Social Informatics. These three forces are: 1) the perceived value of the course in the overall computer science curriculum, 2) the quality of the Social Informatics instructor's teaching, and 3) the number (and frequency of offering) of Social Informatics courses.

5.3 Teaching Social Informatics

This part of the chapter is organized into four sections. In the first section, we outline key issues undergirding the teaching of Social Informatics (see Table 5.1). In the second section, we discuss how these issues may be tailored to emphasize points most germane to specific curricular needs. In the third section, we discuss the importance of developing a critical perspective on computing (and the use of debate to help achieve this). In the last section, we summarize issues relevant to teaching Social Informatics. The goal of this fourth section is to highlight concepts and techniques that should be a part of ICT-oriented curriculum, not to identify a general Social Informatics class. We focus on concepts and techniques for inclusion in a curriculum so that educators can design a number of classes in which these might appear.

5.3.1 Key Social Informatics Ideas

In Table 5.1 we summarize key Social Informatics ideas. These are discussed in the rest of this section.

5.3.1.1 The Context of ICT Use Directly Affects Their Meanings and Roles

As we say throughout this book, context matters. Designing, developing, deploying, and using ICTs are linked to social and organizational dynamics, and these dynamics are situated in specific contexts. This means that an ICT is always linked to its environment of use: It cannot be considered independently from the situation in which it will be used (Kling & Scacchi, 1982; Orlikowski, 1993; Suchman, 2002).

5.3.1.2 ICTs Are Not Value Neutral: Their Use Creates Winners and Losers

Given the contextual nature of ICTs, it follows that they are often designed, implicitly or explicitly, to support social and organizational structures (Kling, 1994). For example, management information systems are primarily designed to support managers, not the system's direct users. Another example is the development of personal digital assistants, which was first predicated on men using them, so that the resulting device was harder for the typical woman to hold (since the female hand is typically smaller than the

Table 5.1 Key Social Informatics Concepts

1. The context of ICT use directly affects the ICT's meanings and roles.

2. ICTs are not value neutral: their use creates winners and losers.

3. ICT use leads to multiple, and often paradoxical, effects.

4. ICT use has moral and ethical aspects and these have social consequences.

5. ICTs are configurable – they are actually collections of distinct components.

6. ICTs follow trajectories and these trajectories often favor the status quo.

7. ICTs co-evolve during design/development/use (before and after implementation).

design premise). This may also be as simple as those with electronic mail being able to communicate more easily than those without.

5.3.1.3 ICT Use Leads to Multiple, and Often Paradoxical, Effects

Because ICTs are contextually dependent, similar ICTs can have different outcomes in different situations (Sawyer & Eschenfelder, 2002). Thus, although many people believe that ICTs will lead to a paperless office or even to increased productivity, ICTs play out differently in practice. Paper use may increase in some places even as it decreases in others; productive efforts may be spent in places where the value added is difficult to assess. These effects may seem contradictory, depending on the level or perspective from which they are viewed. Among the multiple and often paradoxical effects of ICT use are the rise of both intended and unintended consequences (Tenner, 1996). For example, new ICTs were introduced to one department in a local government to improve organizational effectiveness and efficiency. This led to a situation where the work processes of that department's staff soon became enmeshed with the new ICTs. The departmental staff became dependent on the infrastructure to do their work (the intended effect). However, the lack of systematic maintenance and upgrading of this infrastructure led to the ICTs becoming unreliable. This lack of reliability meant that, over time, the office was actually less capable of achieving its mission (an unintended effect).

5.3.1.4 ICT Use Has Ethical Aspects

The contextual nature of ICTs means that development and use raises moral and ethical issues (Eschenfelder, 2004; Friedman & Nissenbaum, 1996; Introna & Nissenbaum, 2000). This set of topics often reflects the most well known of the key Social Informatics issues. Many of these issues, such as an individual's rights concerning privacy in the use of e-mail, are being contested in contemporary society. More subtle issues include the coding and construction of systems that support assessing credit risks (where biases are built into rule bases) or the use of ICTs to remove entire classes of work (and workers) from an organization are not as broadly discussed. ICT-oriented students should be aware that these moral and ethical issues are often confronted as a series of small decisions whose outcomes are unclear and whose links are not easily seen. This leads to where chains of decisions and the value choices that they require are often made implicitly, sequentially, and without an awareness of the larger issues.

5.3.1.5 ICTs Are Configurable

The term "ICT" actually reflects that any ICT-based system is a collection of distinct components. These components—many of which are nearly commodities—are assembled into unique collections for each organization (or social unit, depending on the level of analysis) (Sawyer, 2001). This leads to unique and socio-technical networks. These socio-technical networks, arising from the confluence of social use and similar components, may lead to different technical networks in each social system (see Brown & Duguid, 2000; Sproull & Goodman, 1989). Furthermore, the multiple functions and ability to reprogram (or alter and extend) these functions makes each technical network of ICTs highly re-configurable.

5.3.1.6 ICTs Follow Trajectories

The configurational ability of ICTs is underlain by the trajectories of the components. A trajectory means that any definable component can be seen as an evolving series of products (or versions) (Faraj, Kwon, & Watts, 2004; Fleck, 1994; Quintas, 1994). That is, ICTs have a history and a future. Thus, an ICT-based system's evolution is as much social history as technical progress (Allen, 2004; MacKenzie & Wajcman, 1999). For example, this concept of a trajectory underlies the debate about the functions of Microsoft's Windows and the company's efforts to fully integrate the operating system (Windows) with the browser (Internet Explorer). The U.S. Department of Justice showed that there was no technical reason for this integration. Instead, the U.S. Department of Justice lawyers were able to show that the integration gave Microsoft the ability to dominate the browser market by linking this function to the operating system (where Microsoft enjoys a near monopoly).

Computer programs are being shaped by social structures and political forces. These shaping forces are often both difficult to decipher and hard to anticipate due to the confluence and interaction of so many events. Further, this discussion regarding Windows and Internet Explorer has many parallels with IBM's practice, up until the U.S. Department of Justice intervened in 1968, of enforced bundling of hardware and software. As this example suggests, trajectories tend to favor the status quo.[5]

5.3.1.7 Co-Evolution of ICT System Design/Development/Use

The configurational ability of ICTs also underscores the socio-technical process of ICT design, development, and use that is reflected in every stage of an ICT's life. Projects are selected based on the political and strategic perspectives of decision makers. ICT design reflects an ongoing discourse among developers as well as between developers, the people who will use the ICTs, and other stakeholders. Implementation is a social activity, centered on the re-orientation of work (or life) around a new system (Barley, 1986). A system's use unfolds over time in a form of mutual adaptation between the ICT and the social system into which it has been placed (Leonard-Barton, 1988). This ever-unfolding process, a "design in use," shows the variations in social power (Kling & Iacono, 1984).

5.3.2 Tailoring Social Informatics Concepts for Specific Curricular Purposes

Students involved in an ICT-oriented education reflect a variety of disciplines and each discipline has particular needs relative to Social Informatics concepts. This suggests that Social Informatics concepts must be tailored to meet the curricular needs of that discipline. Comparing the needs of computer science and information systems students highlights a philosophical difference. Most computer science programs are designed (often implicitly) to prepare students for graduate school while most information systems programs are housed in professional schools and are focused on industrial preparation. Given this philosophical difference, the two programs differ in focus. Computer science students' education is often focused on the technical and logical bases of computing. The most relevant Social Informatics issues would be those that help them put these issues in a broader context. Conversely, information systems students' education is focused on understanding and meeting an organization's needs for ICTs. Thus, the focus of their education is on bringing ICTs into organizations to support broad-scale (strategic or operational) needs. Social Informatics concepts can enable students to align ICTs with organizational goals. These same concepts, presented in a different context, can also help computer science students design and develop more useful ICTs. And, as fewer computer science students continue in graduate school, the ability of these programs to provide students with professional preparation becomes ever more important (Gal-Ezer & Harel, 1998; GAO, 2004; Huff & Martin, 1995; NSF, 1996, 2002).

Tailoring the discussion of these concepts for various ICT-oriented students brings practical significance to these issues. Take, for example, the concept of ICTs following technical trajectories. For computer science students, the debate about a current operating system's functionality can only be understood in the light of IBM's move to a common operating system for many of its products, the spread of Unix, and the expansion of simple PC-based DOS support. The same concept of technical trajectory can be illustrated to information systems students by outlining the growth of a market for enterprise resource planning (ERP) systems and the broader move from organizations' building their own software to buying packaged software (Carmel, 1997; Davenport, 1998; Sawyer, 2001).

The tuning of these concepts to reflect issues germane to the various student populations can range from broad-scale (as the previous example of technical trajectories suggests) to nuanced. An example of a more nuanced model of Social Informatics concepts is one in which ICTs are not value neutral and their use creates winners and losers. For computer scientists, the values inherent in ICT decisions show up in subtle ways, like in decisions regarding what information an application should provide to a user. Application designers who choose to pass cryptic messages to users when a program does not operate correctly carry the implicit or explicit belief that the user would not know what to do about such an error. This demands that the people who use it have access to technical help when these cryptic messages appear: having the offending stack overflow data displayed on screen is rarely helpful to the person who sees it. For information systems students, a more comprehensive review of Social Informatics will help them to better understand and anticipate the needs of users, and to dispel the dangerously incorrect stereotype held by many IT professionals that users are technically naive and/or their jobs relatively simple and routine (see, Beath & Orlikowski, 1994; Suchman, 1996).

5.3.3 Social Informatics as Informed Critical Thinking

The concepts of Social Informatics imply a dynamic tension between the positive and negative effects of new ICTs. Perhaps the best way to characterize and act on this dynamic tension is to develop an ability to think critically about the roles and value of ICTs. By critical thinking, we mean here developing in students the ability to examine ICTs from perspectives that do not automatically and often implicitly adopt the goals and beliefs of the groups that commission, design, or implement specific ICTs.

This critical orientation entails developing an ability to reflect on issues at a number of levels and from more than one perspective (Lamb & Sawyer, 2005). This is a difficult and lofty goal, one that is central to most curricular reform efforts in STE&M education (NSF, 1996, 2002). Further, because an informed critical perspective means being able to draw on the research and theories used to develop the findings, this approach implies that high quality research must be synthesized for use by faculty and practitioners in the ICT-oriented communities.

It is important for IT professionals to be able to understand ICT designs, configurations, related social practices, and choices about these practices from multiple perspectives. There are various means to help students analyze the value conflicts and to explore different perspectives in these situations. Faculty who teach these topics have found that having students engage in explicit debates about ICT alternatives is one powerful teaching approach to help them confront ICT choices that they might otherwise ignore or dismiss. Given that the premise of Social Informatics is that social forces help to shape technology, to understand this dynamic requires a discussion of the major social forces involved. These social forces represent multiple perspectives and rarely have clear-cut answers. In fact, focusing on finding answers may obscure the fact that many of these problems have no closed-form solution.

By developing such a critical perspective, ICT-oriented students will be better prepared to contribute to the public debates about the uses and goals of ICTs. The importance of ICTs to the economy and in their broader role as a societal force is constantly being explored by people from many perspectives. In the nascent years of computing, discussions by leading scientists were very influential (e.g., Vanever Bush's "As We May Think" of 1945). More recently these discussions of the social and organizational effects of ICTs have often been polarized. For example, Dertouzos' (1998) bestseller, *What Will Be,* presents a relatively uncritical perspective on the potential roles of ICTs in society, despite its effective portrayal of complex technologies in an understandable form. More commonly, though, discussions regarding the values and uses of ICTs are informal and local—speaking with peers in an office setting—where the individual ICT professional will be an important (and often undervalued) voice.

The goal of bringing ICT-oriented students to a point where they have the ability to draw on techniques to enable reflective, inquiry-oriented analysis of ICT design and use suggests that the Social Informatics material supporting this education be centered on examples, case studies, and student-led projects of local/personal interest. Further, to achieve this perspective building, the concepts of Social Informatics should be introduced early in a technical education. This introduction should be neither compartment (e.g., in a separate class or as an end-of-semester topic in an introductory class) nor deferred (e.g., as an optional elective to be taken in the latter stages of a program of study). Effective presentation of Social Informatics concepts and techniques implies a broad inclusion across the formal curriculum.

5.3.4 Issues with Teaching Social Informatics

We argue that Social Informatics findings and concepts provide students with a set of frameworks to help organize social and organizational forces and to make otherwise seemingly idiosyncratic and odd behaviors more understandable and predictable. Several issues affect the teaching of Social Informatics concepts and the inclusion of these concepts in ICT-oriented curricula:

1. Motivating contemporary ICT-oriented educators to value (and include) Social Informatics concepts and analytic techniques in the curriculum

2. Difficulties with synthesizing Social Informatics literature that is mostly research-based and spread across numerous disciplines

3. Helping students integrate Social Informatics concepts and techniques with their own experiences

4. Dealing with existing mental models that students bring to Social Informatics issues

5.3.4.1 Motivating Contemporary ICT-Oriented Educators to Value (and Include) Social Informatics Concepts and Techniques in the Curriculum

There are three inter-related problems to motivating educators. First, instructors may themselves be unaware of Social Informatics research. Thus, they may not be able to articulate this perspective well or appreciate their role in propagating implicit and explicit norms regarding the approach to the roles of ICTs in either organizations or the larger society. If Social Informatics concepts are presented as anecdotal without a grounding in the empirical research, students are doubly undereducated. These students are robbed of an extensive and insightful body of relevant literature that will assist them in their work, and they are likely to continue developing ICTs that fail to meet the diverse needs of those who use them.

The second problem with including Social Informatics concepts in an ICT-oriented curriculum is the quality of teaching. If the Social Informatics concepts and techniques are seen as secondary, then the teaching of the topics may be relegated to lesser instructors—perhaps a colleague who is seen as a 'jack-of-all-trades' or someone who is willing to do such a course as a service, even though there is no passion (or knowledge or even much interest). This type of devaluing is more insidious than just ignorance. It demeans both the importance of Social Informatics concepts and devalues the hundreds of dedicated faculty who are both passionate regarding Social Informatics research and excited to teach the concepts and techniques to ICT-oriented students.

The third aspect about valuing Social Informatics concepts in ICT-oriented curricula involves the focus of the program of study. Many of the contemporary computer science curricula are structured to prepare students to enter post-graduate programs in computer science, not entry-level ICT-oriented jobs. This attention often downplays the Social Informatics topics. Ultimately this both weakens the program graduate's value to the market and, if current attendance numbers continue, also reduces the number of these students who continue on for graduate work. There is a growing recognition that ICT-oriented programs may need to create separate foci for those heading toward post-graduate education—akin to the differing educations that research-oriented medical doctors have as compared to doctors who are being trained for clinical practice.

5.3.4.2 Difficulties with Synthesizing Social Informatics Literature That Is Mostly Research-Based and Spread Across Numerous Disciplines

The dispersed research literature means that the findings are both difficult to locate and written to meet the needs of the academic publishing outlets where they appear. Furthermore, because the focus is on the social and organizational aspects of technology, excellent lessons driven by outdated technologies are difficult to make clear if students are fixated solely on the value of the ICTs in question. In the absence of textbooks that review the range of research, one intermediate step is to use excellent anthologies regarding the social and organizational aspects of computing (such as George, 2003; Huff & Finholt, 1994; Kling, 1996d; Kraut, et al., 1989). Texts that provide technocentric visions of organizations and society, or that under-emphasize the social aspects (i.e., highlighting only positive outcomes of ICT use), can actually impede the teaching of Social Informatics concepts.

5.3.4.3 Issues with Helping Students Integrate Social Informatics Concepts and Techniques with Their Own Experiences

As we discussed earlier, students may be exposed to important concepts relative to Social Informatics in their sociology, psychology, history, and economics courses as part of their broader studies (Davenport, 1996). However, when these concepts and issues are not brought into the context of ICT design and use, it is difficult for students to sustain such relationships. Thus, discussions of generic impacts, or broad discussions of science and technology impacts (as is often done in introductory courses on this topic) are a partial, and often incomplete, means to provide coherent, grounded discussions of computing's impacts.

5.3.4.4 Dealing with Existing Mental Models That Students Bring to Social Informatics Topics

In the absence of research-based presentations of Social Informatics issues, students will often develop their own—typically naive—conceptualizations. Although students with practical experience may be more responsive to (or at least recognize) the issues, they may have created unstable or poorly defined mental models to deal with Social Informatics interests. For example, Ellen Ullman (1997) writes in *Technophilia*, a book about her work as a software engineer, that she envisioned users of her systems as "… wildebeests of the programming food chain, consumers, roaming perilously far from the machine" (p. 17). Mental models such as these, where the user is so distant and easily discounted, can be difficult to change—even when findings from rigorous empirical research support such re-orderings.

Many of these issues are similar to those present in other STE&M education efforts. For example, physics texts are being criticized for both oversimplifying events and at the same time not being able to bring theoretical concepts into practical use (McDermott, 1993; McDermott & Redish, 1999; Redish, 1996). An example of the latter is that many advanced physics students can describe the formula underlying a particular occurrence but are unable to present a realistic scenario of such motion (McDermott, 1993). This example reflects the same issues that arise in the discussion surrounding user-participation in information systems design.

5.4 Recommendations

In this final section, we present a set of recommendations regarding the inclusion of Social Informatics concepts and analytic techniques into ICT-oriented curricula. This inclusion mirrors the attention being paid by many STE&M disciplines to broadening the value and applicability of the technical aspects of the education by tying the aspects to the larger social context. Our recommendations center on curricular—not course—guidance, because course-level content must be carefully constructed to focus on the particular needs of the many ICT-oriented disciplines.

Recommendation 1: Social Informatics concepts should be integrated into the formal education of ICT-oriented students.

1. Social Informatics content should be seen as an integral aspect of the curriculum. Including Social Informatics concepts also reflects existing accreditation and curricular guides of the ICT-oriented professional societies.

2. Social Informatics concepts and analytic techniques should be sustained across the curriculum. This allows students to develop critical thinking skills to support such analyses.

3. Social Informatics concepts should be introduced early in the curriculum, perhaps in the first class, with a focus on developing these concepts through learning analytic techniques.

4. Social Informatics concepts must be both organized and presented in a way that satisfies the particular curricular emphasis of all ICT-oriented disciplines. This implies that additional attention is needed to develop appropriate pedagogical materials.

5. Social Informatics material should provide a set of organizing principles to help students organize and understand issues.

6. Faculty involved in teaching Social Informatics concepts must be competent with the pedagogies and analytic techniques involved in teaching this material.

Recommendation 2: Social Informatics topics should be taught in ways that develop critical thinking skills.

1. Critical thinking means developing skills as a reflective practitioner, understanding the failure costs, consequences, and risks regarding ICT design/construction and use, having the skills to be both flexible and creative regarding ICT design/construction/use, instilling a sense of inquiry, and developing a problem-centered perspective on ICTs. This last point highlights the policy issues regarding ICTs. This

approach provides students a chance to practice contributing to public debates regarding ICTs.

2. Social Informatics concepts and analytic techniques should use field experience and problem-driven learning to amplify classroom instruction.

In summary, in this chapter we have noted that the ongoing efforts to reform ICT-oriented curricula reflect the ongoing concerns with the lowered interest in STE&M education even as the need for this type of education is growing. This disparity between interest and need is seen in ICT-oriented program graduation rates, most keenly so in computer science. We recommend that academic units that are charged with teaching computer-oriented programs find ways to incorporate Social Informatics material and concepts into the curriculum. Doing so will both increase the value of their students and, more importantly, enable these students to design, construct, implement, and support the use of better (more useful and useable) ICTs. The recommendations also provide a basis for policy makers to set out guidance regarding the curricular and accreditation requirements for ICT-oriented curricular needs. Such guidance will have identifiable outcomes for students, who will learn to use a large body of rigorous empirical work that guides them to better develop ICTs, including analytic techniques and frameworks for orienting and organizing the uses of ICTs in organizations and in our society. These students will more readily realize their potential technical contributions to our organizations and to the broader society in which we live.

Endnotes

1. Much of the discussion in this chapter specifically focuses on these two disciplines. ICT-oriented students also are educated in information science and various social science disciplines (such as communication and education), though we do not have the space in this chapter to discuss specific issues relative to the Organizational and Social Informatics content in the education of these other students.

2. The 2003–2004 Criteria for Accrediting Computing Curricula note for both Computer Science (in standard IV-17) and Information Systems (in standard IV-15) that there be "sufficient coverage of social and ethical issues." For more information about ABET, see http://www.abet.org/cac1.html.

3. There are links to a wide array of Social Informatics courses and degree programs on the Social Informatics Home Page (see http://www.slis. indiana.edu/SI).

4. There are also varieties of courses offered in a range of departments that cover Social Informatics topics that are not included in this discussion due to space constraints.

5. See, for example, the detailed account (Hoffman, 1999) of how word processing software design evolved as developers' notions of the user changed (from designing for female typists to designing for male knowledge workers).

CHAPTER 6

Communicating Social Informatics Research to Professional and Research Communities

The Standish Group has been tracking project outcomes in the USA since 1993. In a recent survey of 280,000 applications projects in large, medium and small cross industry companies, 23% were cancelled before completion or never implemented and 49% exceeded their budgets and time-scales and had fewer features and functions than originally specified (Extreme Chaos 2001, 2001). Some failures are high profile, ... A good many more are never reported: a research survey of 1027 projects discovered that only 130 (barely 13%) were successful.

Drummond and Hodgson (2003)

In a happy sign for technology companies, Forrester Research on Monday said it raised its forecast for U.S. information technology spending growth in 2004 from 4 percent to 5 percent.

"We now expect U.S. IT spending in 2004 to reach $776 billion, and $825 billion in 2005, compared with $739 billion in 2003," Forrester said in a report. ... A Gartner survey of 956 business IT leaders worldwide found that they expect to raise their technology budgets by an average of 1.4 percent in 2004.

Frauenheim (2004)

It is widely known that many large-scale change management projects involving new information technology (IT) fail for reasons unrelated to technical feasibility and reliability.

Markus and Benjamin (1997)

6.1 Learning from Organizational and Social Informatics

The quotes leading off this chapter illustrate the two sides of a problem that organizations regularly face in integrating information and communication technologies (ICTs): In spite of rising corporate investment in ICTs, many large-scale projects continue to fail.

Is each new ICT project so unique that nothing can be learned from the past? Do systems designers and implementers develop ineffective systems because they have an inadequate conception of the people for whom they design these systems (Forsythe, 1992, 1994; Suchman, 1996)? Are ICT professionals so focused on technical requirements that they are not paying attention to academic research on systems success and failures? Markus and Benjamin (1997) argue that one reason for persistent systems failures is social—IT specialists, line managers, and others involved in the change effort hold a "magic bullet theory" where they believe that "IT itself has the power to create organizational change." McKeen and Smith (1997, p. 17) agree and, coloring their bullet silver, state that "All too often IS managers have been guilty of looking for a silver bullet from technology. The one magic answer that will solve all of their problems."

The costs of this belief are immense and growing. Dalcher and Drevin (2003, p. 137) estimate that "billions of dollars are wasted each year on failed projects." Many expensive information systems are underutilized, not producing their promised value, or are outright failures, yet organizations allocate increasing percentages of their operating budgets to expenditures on information systems. According to Carr (2004, p. 28) "nearly every sizable company has its own horror stories about IT projects that went wildly over budget or off schedule, that never came close to delivering the promised benefits, or that were abandoned." So long as ICT professionals continue to think of ICT as a magic bullet, they have few incentives to turn their attention to the findings and insights that have come from the careful academic study of both successful large scale change management projects and system failures. The critique of the technological determinism on which this "magic bullet" perspective is based has been in the literature of information systems for some time (and it is raised again in Chapter 2), yet many ICT professionals have not gotten the message. What can be done to disabuse them of the notion that the implementation of ICTs is sufficient to cause organizational change? What can be done to broaden their perspective on varied roles ICTs can play in social and organizational change?

In this chapter we ask Social Informatics (SI) researchers to shoulder the responsibility for communicating the core of Social Informatics (defined here as its assumptions, concepts, theories, insights, and findings) to ICT professionals and other academic research communities. It begins with a brief description of these two groups, describing them as the main audiences for SI outreach efforts. After a discussion of some of the challenges involved in attempting to reach out to these audiences, several strategies are offered as ways to remove the barriers to and improve the flow of information between SI researchers and these audiences. Some of these strategies are low cost in terms of time and effort and can begin immediately. Others are more difficult and will require more concerted collective action to accomplish. The responsibility for outreach activities is great, but the stakes are high—organizations must learn to manage their growing investment in ICTs more effectively (Carr, 2004). This requires a change in the ways in which ICT professionals think about technologies, organizations, and social change.

6.2 Audience

Social Informatics research should be communicated to three distinct groups: ICT professionals, researchers and teachers in a range of academic disciplines, and policy makers—particularly those involved in funding research into the social and organizational aspects of computing. Reaching ICT professionals is of primary importance because they are responsible for managing their organizations' investments in ICTs (Addison & Vallabh, 2002, p. 128).

> As business reliance on software grows, so do the business-related consequences of software failure. Most software projects take place in an unpredictable environment in which many pitfalls exist that may affect the successful outcome of a project. Post-project evaluations reveal that many of the problems encountered were in fact predictable.

Many ICT professionals have an inadequate understanding of ICTs, the actions and interactions of the people who use them, and the organizational and social contexts in which they are used. Paradoxically, this is a rapidly growing group of professionals that can be expected to have a major impact on the ways in which ICTs are designed, managed, and used in the coming years. As we have outlined in Chapters 2, 3, and 4, a major goal of SI is to develop theories and produce reliable, empirically based knowledge that can help those who design, implement, and manage ICTs by enabling them to improve the lives and work of the people who routinely use these technologies. The insights of SI can be used to broaden and deepen their conceptions of the relationship between ICTs and organizational change. This chapter will describe a set of strategies that can be used to communicate to this audience.

Communicating SI research to other academics, both researchers and teachers, is important because the value of SI theory, insights, and findings transcends ICT-oriented curricula and has relevance across a range of disciplines. The challenge is to draw SI work together and begin to make it known to other academic research communities in the United States and abroad. This is difficult because SI researchers are distributed across a wide range of academic disciplines (Kling, 1999a).

We estimate that there are about 200 people in the United States doing SI research and probably another 100 in Western Europe, Israel, Japan, South Korea, and Australia. They are working in such diverse fields as information science, computer science, information systems, sociology, anthropology, journalism, and communications. There is a need to educate a wide group of academics about the work of these researchers. An understanding of the research can enrich their teaching, and will begin the work of legitimating and raising the profile of SI. This chapter will describe several strategies that can be used to gather SI work from these diverse sources and make it accessible to the wider academic community.

In addition, as is argued in Chapter 5, many ICT-oriented disciplines do recognize the significance of SI research, but it is not given adequate treatment in many classes. Better communication can increase the familiarity that other academics have with SI theory and research, improving their own teaching and research. This is particularly important for those involved in training the next generation of ICT professionals. In Chapter 5 we lay out the case for including the core concepts, frameworks, and research exemplars of SI in the undergraduate and graduate curricula of ICT-oriented disciplines. In this chapter, the emphasis is on communicating the research to other academics outside of classrooms and suggesting strategies that are associated with discipline building such as hosting workshops and conferences and producing publications.

Communicating SI research to potential funders of research is also important. This community should have a familiarity with the core concepts, questions, frameworks, and research exemplars of SI so that they can make informed decisions about supporting research proposals that employ this approach. Finally, engaging in discourse about SI research with policy communities is essential if there is to be continued and broad-based funding and other support for further research. As policy analysts and decision makers become familiar with the basic themes, concepts, and findings of SI, they can make better decisions about funding research into the design, development, procurement, deployment, management, and use of ICTs. This chapter will discuss strategies that can be used to reach the ICT professional, academic research, and policy analyst communities.

6.3 Communicating to ICT Professional Audiences

Which ICT professionals could most benefit from an awareness and knowledge of relevant SI research and findings? A partial list of occupations includes those who design, implement, manage, and work on ICTs; those who make decisions about ICT development, procurement, and deployment; and those for whom ICTs assume a central role in their work.

ICT professionals include:

- ICT project managers
- Software and hardware designers
- ICT consultants
- Project managers
- Middle managers
- Executives
- ICT system designers
- Quality and usability testers
- Information managers
- Systems analysts
- Knowledge workers (e.g., corporate librarians)
- Recruiters and human resource managers

Why is it important to communicate with this community about the role of social and organizational contexts in the design, implementation, and use of ICTs? One answer is that the incorporation of key ideas and findings from SI research on ICTs can improve professional practice. Managers, analysts, and consultants can benefit from a deeper understanding of how people and organizations work with ICTs. They can benefit from this research because they gain more sophisticated understandings of the effects of the social and organizational contexts on ICT use, and of the ways in which people actually use these technologies. ICT developers can make use of this research to develop more efficient, usable, and useful information systems that can be smoothly integrated into the social and organizational settings for which they are designed. By becoming familiar and keeping current with research that focuses on ICTs and social and organizational change, they can improve the systems they design and implement, making them more workable for the people who use them (Kling, 1999a).

Costly systems fail, under perform, or do not return the value of their investment for reasons other than their technical configuration of feasibility (Addison & Vallabh, 2002; Carr, 2004; Markus and Benjamin, 1997). Another answer is that greater understanding and awareness of SI research can help ICT professionals avoid implementation failures. When information systems projects fail, there are direct and indirect consequences for the organization. For example, in addition to economic costs, there are social costs (Donaldson, 2004). Those involved in designing and implementing a failed system may feel alienated or may experience lowered morale. Those who attempt to use the system may experience strained relations with the ICT professionals involved in the project. The decision makers in the organization may have less trust in those who champion the project and in the vendors who supply the system components.

Turning to SI research, ICT professionals can take advantage of analysts who systematically and rigorously study how people work with information systems in a range of organizational settings. For example, they can learn that systems fail because:

- Designers lack an understanding of the current situations of those who will use the new systems (Donaldson, 2004). How locked-in are people to their current systems? What are the switching costs involved in changing over to a new system?

- Designers lack an understanding of the organizational setting in which the systems will be used (Heeks, 2002).

- The systems have not been adequately designed for those who will be using them. They do not have the types of features and functionality that is needed for the tasks at hand and cannot easily be integrated into the existing work flow (Addison & Vallabh, 2002; Roberts & Bea, 2001).

- The costs of learning the new systems outweigh the benefits that workers can get from using them. One factor that can tip workers against the adoption of new systems is increased complexity that does not carry a clear payoff.

- The design process has not taken into account what people actually do (Wilson & Howcroft, 2002). A system that is developed to handle a well defined set of tasks may not be able to cope with the unstructured and fluid portion of a worker's job. In many cases, this may claim a significant amount of the worker's time.

Understanding this type of research does not guarantee the success of every ICT implementation project. However, it does provide ICT professionals with the type of knowledge that can improve their design, development, and implementation processes by placing the technical design specifications of the new system in a broader social and organizational context. This allows them to anticipate the types of problems that can arise when people are being asked to move to new information systems. It also encourages them to stretch the time horizons of projects to be able to observe the systems in use. Awareness of these types of SI research and findings also allows them to challenge the assumptions they have about the people who will be using the systems they are designing. For example, by taking into account what people actually do when they are using a current system, ICT professionals can ask questions about what the new system will do and how people are likely to react to it. How much of a change will people be asked to go through as a consequence of the implementation of the new system? What will people have to do to integrate the new system into their work? How will their work be changed? What are some ways that they may reconfigure the ICT as they work with it?

It is especially important to reach this group because the number of ICT professionals is expected to grow rapidly between 2002 and 2012. According to projections from the Bureau of Labor Statistics (2004a), jobs on the information sector are expected to increase by 18.5 percent (annual growth rate of 1.7 percent), making it the fourth fastest growing industry group within the service sector, which itself is the fastest growing part of the economy over this time period. Three of the top ten fastest growing industries are ICT-related: software publishers (69 percent increase in employment, 5.3 percent annual growth rate), computer systems design and related services (55 percent, 4.5 percent), and Internet services, data processing, and other information services (46 percent, 3.9 percent). Out of the ten fastest growing occupations, four are ICT related: network systems and data communications analysts (57 percent increase by 2012); medical records and health information technicians (47 percent); computer software engineers, applications (46 percent); computer software engineers, systems software (45 percent). As a consequence of this job growth, there is increasing demand for ICT management (Bureau of Labor Statistics, 2004b):

Employment of computer and information systems managers is expected to grow much faster than the average for all occupations through the year 2012. Technological advancements will boost the employment of computer-related workers; as a result, the demand for managers to direct these workers also will increase. In addition, job openings will result from the need to replace managers who retire or move into other occupations. Opportunities for obtaining a management position will be best for workers possessing an MBA with technology as a core component, or a management information systems degree, advanced technical knowledge, and strong communication and administrative skills.

Despite the recent downturn in the economy, especially in technology-related sectors, the outlook for computer and information systems managers remains strong. In order to remain competitive, firms will continue to install sophisticated computer networks and set up more complex Internet and intranet sites. Keeping a computer network running smoothly is essential to almost every organization. Firms will be more willing to hire managers who can accomplish that.

These professionals can be expected to have a profound effect on how ICTs will be designed, implemented, and used in a wide range of organizational settings. They will have learned a set of technical skills in their formal education, on the job, and in career development courses. Hopefully, they will have acquired the ability to learn how to learn, so that they will be able to keep their skill sets current. For those whose work is many steps removed from the front lines of ICT use, technical skills may be all that are needed. However, there is a growing awareness among educators and employers that a technical skills set is necessary but not sufficient for ICT professionals who are closer to those who use systems routinely in their work.

For these people, computing and systems expertise must be augmented by knowledge of the organizational and social impacts of computerization. This knowledge is not typically found among graduates of information systems, information science, and computer science programs (Huff & Martin, 1995; Mann, 2002. By implication, it is also not present to any significant extent among many in the current ICT workforce. One of the values of SI research is to provide insights and findings that can be used by current and future ICT professionals to improve people's' abilities to work effectively with ICTs. As the costs of ICTs increase and the systems designed around them become more complex, the pressures on ICT developers and managers to implement successful systems will also increase and this knowledge will increase in value.

Communicating SI research effectively to ICT professionals means influencing two crucial components of their worldview. The first is how they conceptualize processes of software, hardware, and systems development; implementation; and use. The second is how they understand the complex relationship between ICTs and the social and organizational contexts in which they are embedded. Familiarity with SI research turns ICT professionals' attention to the social and organizational contexts of ICTs, the mutual shaping of technology and its contexts, and the configurability of ICTs, broadening their world view and enabling them to improve their practices. In Chapter 5, we argue for the critical importance of reaching the next generation of ICT professionals by integrating SI concerns into graduate and undergraduate curricula. Here, the focus is on reaching the current generation.

There is another reason why the effort to communicate with this audience is worthwhile. There has been a long history of support for academic research on ICTs from the professional community, and this support—both in terms of funding for research and in the provision of access—must continue. Westfall (1999) makes the point cogently, stating that "As an applied field, we need financial support and institutional validation from practitioners." Although Westfall is addressing researchers in information systems, this point can be generalized to SI researchers as well. In order to maintain this type of relationship with ICT professionals, it is incumbent upon SI researchers to convince practitioners of the worth of their research. They must make clear the value of understanding the powerful ways in which ICTs shape organizational and other contexts and are, in turn, affected by these contexts, and demonstrate the ways in which the design, development, and implementation of ICTs affect and change work. They must impress upon the community the importance of continued support for research into the organizational and social impacts of computerization.

There are two challenges that must be addressed when communicating SI research to ICT professionals. The first is perceptual: The research must be seen by this group as relevant, timely, and valuable and must contain suggestions that can be realistically implemented in their practices. The second is competitive: Academic SI researchers are not the only group that is "selling" research to ICT professionals (Davenport, 1997).

6.3.1 Perceptions of the Relevance of Social Informatics Research

Some perceptual barriers are that:

- Potentially relevant research is not easily accessible to ICT professionals.

- The ICTs with which this audience works are not the focus of much research.

The first challenge in reaching this audience is to overcome two barriers that seem to exist between ICT professionals and academic SI researchers, both of which contribute to the perceptions of the former that much academic research on ICTs is not worthy of their attention. These barriers are rooted in the academic research process; the first is related to the conduct of research and the second arises from the choices made by researchers about what they will study. As is detailed in Chapter 5, many ICT professionals currently believe that research from the academic community is largely irrelevant and that attention to the social contexts of ICT design, implementation, management, and use is not important to their work (Agre, 1996; Lang, 2003; Senn, 1998). The problem is summarized neatly by Lang (2003), who argues that "unfortunately, much of what IS academics have to say never reaches the ears of practitioners for a variety of reasons." For example:

- [The] Society of Information Management International (SIM), whose membership is primarily CIOs, decided in 1995 to stop bundling MIS Quarterly with membership. Few members opted to continue their subscription even at a discounted price, and non-academic subscriptions declined by more than 60 percent (Westfall, 1999).

- In an interview, a CIO stated that "The work is not relevant, readable, or reachable" (Senn, 1998, pp. 23–24).

- "Few IT academics are viewed as the world's authorities on IT in business; we are seldom sought out for our opinions on contemporary IT issues. The journals in which academic IT research is published are rarely read by practitioners" (Davenport, 1997).

These examples are supported anecdotally by Kling (1996b, p. 34), who comments that:

> Relatively few of the practicing technologists whom I have met in some of the nation's top industrial and service firms read broadly about the social aspects of computerization. Sadly, a substantial fraction of computer specialists focus their professional reading on technical handbooks, such as the manuals for specific

equipment, and on occasional articles in the computer trade press.

Why do such perceptions and patterns persist? Why don't ICT professionals read academic research, especially articles containing social analyses of computing in the types of organizations in which they work? What could make sound, reliable, and potentially valuable academic research appear to be irrelevant and unreadable? What barriers come between researchers and professionals that reinforce the latter perceptions that the research lacks applicability to their workplaces? One barrier may be rooted in the process of academic research, which poses a dilemma for researchers wishing to reach the ICT professional community. To create and maintain a stream of research may take three to four years because of all of the articulation work that must be done. There are proposals to be written, grants to be sought, access to be gained to research sites, long periods of data collection and analysis, and, finally, articles and research reports to be written. An important goal of this process is to generate reliable knowledge about ICTs and social and organizational change. At times during this process, researchers are faced with a choice—when should research be presented to peers and when should it be presented to practitioners? The dilemma here is that there are different criteria that influence the writing for each audience.

Researchers have to follow one set of norms to publish research for their peers and another to publish research for a nonacademic audience. To publish in peer-reviewed journals and create the type of research output recognized as a standard "scientific article," academic researchers must report research in a style that conforms to a widely accepted set of standards. This involves the use of a particular structure, writing style, and logic of argumentation. Characteristics of this format may contribute to the perception among practitioners that much academic research is difficult to read and irrelevant. For the academic audience, however, these same characteristics are indicators of the reliability of the research. For example, two sections of an academic research report that are required for peer-reviewed publication are the literature review and description of the methodology. Both of these sections may make research articles more difficult for a nonacademic to read, because they do not contain descriptions of findings or pragmatic suggestions for improving practice.

To publish in a trade or industry journal read by ICT professionals requires a more journalistic writing style. There is more emphasis in this type of writing on accessibility, on the explicit linking of the research findings to improvements in the professional's practice. Vivid and compelling examples are an important feature of this type of article. The standard structure used when writing an academic research publication is not the standard in this domain. There will not be a lengthy problem statement or intricate discussion of the strategies used to control error or to maximize reliability and validity. The emphasis is on the pragmatic implications of the research. Articles typically conclude with suggestions, strategies, or tactics that can be used in the workplace to improve some aspect of organizational performance or structure.

Clearly, there are some important differences between the type of research report that can pass through academic peer review and that which can withstand the critical eye of an editor of a practitioner-oriented publication. Writing for an academic audience emphasizes reliability; articles for ICT professionals emphasize accessibility. When faced with the choice of pursuing the scholarly or practitioner-oriented publication of research, many researchers decide to write research reports that conform to the norms of formal scholarly communication. There are many good reasons for pursuing this option (Moody, 2000). It is an important and understandable choice. Scholarly publication is an essential step in advancing both the researcher's individual stream of research and the larger body of reliable and accepted research in the academic domain in which the researcher works. Peer-reviewed articles are important in the academic reward system and writing for practitioners becomes a lower priority for many researchers. As a consequence, ICT professionals do not find many research articles to be reader friendly (Lang, 2003). Such a reaction is sensible, given the realization that the work is written for an academic audience and not for them. In order to overcome this barrier, it may be necessary to provide career-advancing incentives to encourage researchers to write for ICT professionals.

A second barrier that may contribute to the perception among ICT professionals of the irrelevance of academic research is a consequence of the choices made by some researchers who study ICTs in organizations. According to Attewell (1998), some researchers get caught up in the tidal wave of rapid technological change in ICTs and face "a constant temptation to turn away from studies of current outcomes of existing information technologies, and instead turn toward a kind of futurology or speculative stance about what might be the case in the future." Those who have given in to this temptation, he argues, have produced research that has had "unfortunate implications." For example, this type of work typically has an emphasis on speculative theoretical models rather than on careful empirical studies of existing systems ("ought to be" instead of "is").

Some researchers choose to focus on the most innovative ICT applications in the most forward-looking and dynamic organizations instead of on the routine uses of ICTs in average organizations, "the point being that what one observes in the largest, most resource-rich, and most committed settings is not a good predictor of the typical effects of a technology in the larger world" (Attewell, 1998). Attewell also points toward a tendency to minimize the importance of findings that indicate problems with ICT design, implementation, and use by attributing the effects to the "beta" versions of systems. This tendency is a form of technological determinism because it attributes the presence of problems to the direct effects of ICTs. The implied assumption is that when later versions of the system are implemented, these problems will disappear, having been designed out of the system. Much SI research can act as a corrective to this tendency if it can be brought to the attention of ICT professionals. As discussed in Chapter 3, there is a stream of research that for two decades has been demonstrating that approximately "40 percent of systems projects in major corporations are total failures," in large part because of a lack of attention to social and organizational factors.

There is a belief among ICT professionals that academic research on ICTs is not relevant. Two factors that contribute to this perception are the academic style of writing that is a normal outcome of the research process and the focus of some IT researchers on more innovative applications of ICTs. Articles that are written for academic audiences emphasize the reliability of the research in ways that make the articles less accessible to professional audiences. If the ICTs being studied are too far in advance of the types of ICTs that form the installed base in many organizations, the professional audience may have a hard time extracting the findings and insights that will be of practical value to them. This barrier must be overcome if researchers are to have a chance to change the perceptions of ICT professionals and make them aware of the important and potentially relevant findings of SI research.

6.3.2 Competition for the Attention of the ICT Professional Audience

The second challenge in reaching ICT professionals is due to the fact that SI researchers have not cornered the market on ICT research. In fact, there is strong competition from ICT research and consulting companies for "mind share," or the collective attention of ICT professionals. Many ICT professionals and their organizations are in the interesting position of being more willing to purchase research reports from these companies at high prices than to cull relevant research findings from the academic literature. There are some good reasons for this. Searching for, retrieving, evaluating, and repackaging academic findings involves greater opportunity costs, in terms of time and money, than most ICT professionals are willing to pay. As mentioned earlier, the research is difficult to read, generalized, and not easily applicable to the specific problems of specific organizations. Extracting practical insights that can be used to improve practice is not a trivial task.

ICT research and consulting firms have an advantage because they can work with academic research more easily than can ICT professionals and can interpret and package it in ways that address a client organization's specific needs. Because the consultant's report is written expressly for the client using language that is native to ICT professionals, it appears to be more readable and relevant that a standard academic research report. SI researchers are at a disadvantage when it comes to the ability to present research in "practitioner-friendly" packages.

ICT research and consulting firms can also bring many more resources (human, material, and economic) to bear on a problem than can most academic researchers. Teams of consultants and analysts can be dedicated to specific projects on a full-time basis, allowing them to work more quickly than can academic researchers. Their research products have to satisfy internal criteria of quality and do not have to go through external review. This means that they can conduct research with a much faster turnaround time than is typical of academic researchers, who have to put their work through lengthy processes of peer review and revision. Although this is one way of ensuring that published work conforms to academic standards of quality, it also means that many months, and sometimes years, can elapse between the submission of a research report and its publication. SI researchers are at a disadvantage when it comes to timeliness.

Consultants typically focus their research much more narrowly on the specific problems of the organization for which they are consulting with a goal of developing specific resolutions to these problems. Their success depends on their ability to demonstrate to ICT professionals that their work can resolve the problems they have investigated in cost-effective and timely ways. Most academic researchers studying ICTs in organizations are not motivated by these same goals. Where research and consulting firms are conducting research to develop pragmatic resolutions to problems with ICTs in specific organizations, academic researchers tend to pose broader research questions, looking for patterns among organizations. Academic researchers may also use data collection techniques, such as ethnography or participant observation, that are much more time- and data-intensive. Organizational informatics and SI researchers are at a disadvantage when it comes to allocation of resources for research and the pin-point focus that generates research relevant to specific organizations. Research and consulting companies therefore present a significant challenge to academic researchers attempting to wrest some "intellectual market share" out of this marketplace.

SI researchers face two main challenges when communicating their research to the ICT professional community. They must develop strategies that will enable them to overcome the perception that their work has little relevance for practitioners, is not timely, and is not focused on the "real problems of real organizations." They also have to develop strategies that will allow them to draw the sustained attention of ICT professionals, whether this means competing or cooperating with ICT research and consulting firms. The following section outlines six strategies that address these challenges.

6.3.3 Strategies for Communicating to ICT Professional Audiences

Six strategies to improve communication between SI researchers and ICT professionals are:

1. Develop a current empirical assessment of what we know about the worlds of ICT professionals.

2. Conduct and communicate research that is useful for these professionals given what we know about their typical problems, concerns, and information behaviors.

3. Focus on speaking to and writing for ICT professionals.

4. Hold regular forums that bring academics together with ICT professionals.

5. Influence current professional practice through workshops, seminars, and life-long learning.

6. Create "ICT extension services."

To manage competition with research and consultant firms, SI researchers can:

-   Distinguish their work from that done by these firms, emphasizing the value that can be added to practice from academic work.

-   Explore ways to cooperate with these firms.

What can be done to improve and extend vibrant and stable two-way channels of communication between SI researchers and the ICT professional audience? What are some reasonable strategies that SI researchers can use to communicate their work to ICT professionals in ways that positively affect professional practice? What can be done to change the perception among ICT professionals that research on ICTs is difficult to read and seemingly irrelevant? How can the competition with ICT research and consulting firms be managed so that their market share is not impacted, and the insights and findings of SI research can be brought to the attention of ICT professionals?

This section addresses these questions with a set of six broad and related strategies that may provide researchers with ways to get their work into the hands of the ICT professional audience. The first strategy is to develop a current empirical assessment of what we know about the work worlds of ICT professionals. The second is to conduct and communicate research that is useful for these professionals given what we know about their work, their typical problems, concerns, and information behaviors. The third is based on publicity and focuses on speaking to and writing for ICT professionals. The fourth strategy is learning from ICT professionals by bringing them together with SI researchers in regular forums. The fifth strategy is to influence current professional practice through workshops, seminars, and life-long learning. The sixth strategy is to support the development of "ICT extension services," based on the model of agricultural extension services.

The long term goal of this effort is to persuade this community that they can turn to academic SI researchers for high quality and understandable research findings and insights that will be useful in their work. Stated more strongly, the professional ICT audience should look routinely to SI researchers for ways to improve their practice and deepen their understandings of the complex interrelationships among ITCs, the people who work with them, and the organizations in which they work.

6.3.3.1 Learning About ICT Professionals

What do we know about the worlds and information behaviors of ICT professionals? How do they seek and use research on ICTs?

The first strategy is to develop a current and reliable understanding of the social and organizational contexts of ICT professionals. To improve and extend lively and ongoing channels of communication with this audience, SI researchers must learn how to communicate with ICT professionals, instead of expecting that this audience will change its habits to accommodate the constraints of academic research. To do this, researchers should have timely empirical knowledge about what ICT professionals do and how they work, particularly in terms of their information seeking and use. This requires an understanding of the organizational information environments in which ICT professionals routinely operate and the range of information behaviors they enact in these environments. Initial research might segment ICT professionals into the different groups that make up this audience. Some questions to ask about these different groups might include:

- What are the main channels through which ICT professionals obtain information about research?

- What are their patterns of environmental scanning?

- What types of research about ICTs are they most likely to seek out?

- At what points in their workflow and for what purposes are they most likely to use research?

- In what forms should research be presented to them to maximize the chances that they will read and use it?

- What are the typical time frames of these professionals as they engage in information seeking and decision making about ICTs?

Answering these questions is a step toward being able to use SI research to provide ICT professionals with the research information they need when it is needed, in the forms in which it is needed. This type of empirically based background information can be used to develop strategies for communication that can be much more precisely developed for this audience.

Some of these questions can be answered by examining state of the art literature reviews of ICT managers and professionals and their uses of information about ICTs, and, if necessary, conducting new reviews (see Katzer & Fletcher, 1992). For example, there is a significant body of research about the information behaviors of designers, developers, managers, and executives to understand how these organizational members seek and use information through environmental scanning. According to Choo (2001), people in organizations:

> ... scan the environment in order to understand the external forces of change so that they may develop effective responses which secure or improve their position in the future. They scan in order to avoid surprises, identify threats and opportunities, gain competitive advantage, and improve long-term and short-term planning. ... Environmental scanning includes both looking at information (viewing) and looking for information (searching). It could range from a casual conversation at the lunch table or a chance observation of an angry customer, to a formal market research programme or a scenario planning exercise.

Another important facet of organizational information seeking and use is the role of research information in organizational decision making. This is one part of the ICT professional's work where SI researchers would hope to make a difference. For example, when ICT managers are in the midst of a process of purchasing ICTs, such as when upgrading the computing in a call center, at what points in the decision making process does research information make a difference? How is this information sought out, evaluated, and used? In what forms is this information most likely to have an effect on the manager? Some research is not comforting—managers will use information in a post-hoc way to support and justify decisions that they have already made (Meltsner, 1976). They will not pay attention to the work of their own analysts and will instead make decisions "on the basis of politics and personal loyalties rather than the information that the analysts have to offer" (Feldman, 1989, p. 93). However, there is a need for analyses of recent research on the uses of information in organizations to check on these findings, and to assess the changes that have occurred as organizations make use of new ICTs in their routine business and communication processes. Such analyses would provide a contemporary description of ICT professionals' information seeking, evaluation, and use behaviors, giving SI researchers a better understanding of how, when, and in what forms they can communicate their research to this community. It is also possible that new research may be needed, especially if ICT professionals' use of research information is shown to fall into the domain of "artful integrations," defined by Suchman (1996, p. 407) as a concept intended to:

> Draw attention to aspects of systems development and use that have been hidden, or at least positioned in the background or shadows, and to bring them forward into the light. These include various forms of professional configuration and customization work, as well as an open horizon of mundane activities involved in incorporating technologies into everyday working practices, and keeping them working.

6.3.3.2 Redesigning the Research Focus

How can SI researchers begin to develop compelling examples from their studies that illustrate the social embeddedness of ICTs? How can they demonstrate that the mutual shaping of ICTs and their settings helps to explain instances of successful system design and implementation and of large scale system failures?

A second strategy is to redesign the approach used when conducting the type of research that is intended to address problems and issues of importance to ICT professionals and generate findings to be communicated to this audience. The purpose of this redesign would be to produce research that has the qualities (relevance, timeliness, and readability) that this audience seeks in the literature it is likely to read and use. This strategy should be based on the outcome of the first strategy, which should produce a set of problems and issues that are relevant to ICT professionals and engage the interest of SI researchers. In a paper on improving research on information technology, Attewell (1998) suggests that researchers "should pursue empirical studies of existing technologies in real settings, as distinct from speculative or purely theoretical exercises," and that "care should be taken to include representative organizations/settings, not just cutting-edge or high-tech ones." Truex (2001, pp. 4–5) adds to this insight the argument that typical IS research choices reflect a class bias; these researchers operate with a basic assumption "of the right of supremacy and the inherent goodness of management" and argues that this bias "privileges the opinions and positions of managers, professionals, and those in control of organizational power structures." He argues that (2001, p. 6):

> Perhaps in studying our objects of inquiry from the bottom up might yield some surprising insights. Labor unions, community centres and community organizations serving the poor and disenfranchised and powerless need the insights and the skills we have to offer. Moreover, in listening to the voices of these constituencies we learn a lot about the impacts of the systems we construct. It challenges our values, preconceptions and sharpens our research senses.

These prescriptions direct researchers toward more typical organizational settings within which they should make social and organizational contextual variables central to the analysis of computerization, increasing the chances that their findings will have more relevance to ICT professionals. An interesting question is whether the "product to market cycle" can be shortened without sacrificing the quality of SI research intended to influence ICT professionals. Should academic institutions create small, flexible research groups that focus on ICT research for the professional audience?

By focusing on the fundamental themes of SI, as described in detail in Chapter 3 and presented in tabular form in Chapter 5, researchers can begin to develop compelling examples from their studies that illustrate the social embeddedness of ICTs. They will be able to demonstrate that the mutual shaping of ICTs and their settings helps to explain instances of successful system design and implementation and of large scale system failures. For example, a National Research Council report (CSTB, 1998) describes a study by anthropologist Julian Orr, who conducted fieldwork among Xerox service technicians and found that they relied on shared narratives instead of company literature to solve problems in the field.

System designers took this information and used it to develop a "community-validated tips database" which, upon implementation, was accessed more than 1,000 times daily. The importance of attention to the social and organizational contexts in the successful deployment of this information system is described in the CSTB report (1998):

> Both developers and managers attribute the success of this system in part to the effort to take seriously social science ideas about community. They learned that local knowledge conveyed in the community vernacular by community members is useful to technicians troubleshooting unfamiliar problems. And so the system was designed to support vernacular content. They learned that community knowledge could spread much more rapidly than standard corporate publication or validation cycles. And so the system was designed to include many human validators in order to ensure very short validation cycles. Initially, developers and managers worried that they might have to provide economic incentives for technicians to contribute tips. But they learned that technicians value the social validation that comes from other community members who appreciate their tips.

Research attention to the unintended consequences of ICT design, implementation, and use also can provide the types of findings that can help ICT professionals understand the importance of the configurability of the technologies with which they work.

An intriguing possibility for redesigning the approach to SI research that is intended to influence ICT professionals is to create research-based partnerships with organizations in which they work and involve them in the research. Researchers can engage in action research where SI insights are put into practice in organizations during the investigation of some problem (see Truex, 2001). One goal of this work would be to publicize the results of the research, indicating through empirical evidence and the support of the ICT professionals in the organization, that there were gains and benefits that accrued to the organization as a consequence of the research. For example, Sneyd and Rowley (2003, p. 42) describe a project that took place in a large pharmaceutical company, where the objective was to link strategic objectives and operational performance; with support from senior management, they collaborated with four work teams and, reflecting on the beginning stages of the project, comment that the:

> ... early workshops provided an action learning experience for all those staff involved, including the researcher, who switched from being "scorekeeper" to being "coach." By taking staff through these workshops the researcher was able to generate enthusiasm, commitment and action from the participants. These were essential pre-requisites for the continuous success of the research project. ... Throughout the pilot phase and during the structured workshops, the staff debated and discussed which performance measures were important to capture in order to monitor their work processes, and for them to improve their level of service to their customers. These performance measures were linked to overall strategic objectives. This linking of tactical objectives with the more strategic objectives was a positive way to bring about commitment and support for the use of perform-ance measures. Unlike many past initiatives, this project was not just a "flavor of the month." On the contrary, it formed the basis of a new style of management and new methods of working throughout the whole of the analytical development group. (ADG)

If ICT professionals are brought into the research process at early stages, there is a greater chance that they will provide more support for the study, especially if there will be clear benefits to their organization as a conse-quence of the research. They can also help to sharpen the focus of the prob-lem statement, research questions, or hypotheses because they have deep local knowledge. As they become more invested in the research, they will be more likely to find the results relevant to their situations. This type of partic-ipatory research, however, is more likely to influence local practice in the organizations that are studied. The challenge is to leverage the support that should be forthcoming from the local ICT professionals into acceptance of the findings from their peers at other organizations.

6.3.3.3 Publicizing Social Informatics Research to the ICT Professional Audience

How can SI researchers bring their work to the attention of ICT professionals?
A third strategy involves writing for and speaking to ICT professionals about the findings and insights of SI research. One of the products of the first strategy should be a listing of the publication outlets that are frequently used by ICT professionals when they are searching for and reading research-based information. The list should be complemented by an analysis of the forms in which research information is most likely to gain and hold the attention of

ICT professionals. SI researchers should then develop versions of their research reports for these publication outlets.

The articles should be written in a journalistic style, and emphasize the research findings in pragmatic ways that can influence the practice of ICT professionals. Reports of system failures can describe the chains of events that led to the failure so that readers can learn what to avoid or how to recognize the early warning signs. Problem-driven reports can be prepared that present useful data generated by SI researchers about issues and problems of concern to ICT professionals. These reports could include survey results, illustrative case studies, meta-analyses of ICT research, and syntheses of research findings. There are examples of "crossover" academics who write effectively for this audience, see, for example, Kanter (2004) and Wegner, McDermott, and Snyder (2002). There are also publishing opportunities for SI researchers to write books intended to "inform the practice and management of information systems"; for example, the Wiley Series in Information Systems seeks monographs that explicitly address issues of practice. In addition, SI researchers should consider presenting their work at the conferences that ICT professionals attend.

If this strategy is to succeed, there must be some type of reward or recognition within the academy for SI researchers who choose to write for and present to the ICT professional audience (Lang, 2003). They are choosing a riskier career option compared to a traditional academic researcher's career trajectory. The time they spend reaching this audience means that they are producing fewer peer-reviewed research articles and, if they wish to publish their research in traditional scholarly journals, they are actually writing second versions of their articles for practitioners. There should also be institutional support for researchers who present their work at practitioner-oriented conferences.

6.3.3.4 Holding Regular Forums That Bring Academics Together with ICT Professionals

What can be done to bring academics and ICT professionals together in a setting where the practitioners can educate the academics about the issues and challenges of importance to them?

Another way to communicate with ICT professionals is to solicit their input on issues and challenges that are important to them. SI researchers can invite managers, designers, and other ICT professionals to attend meetings. These practitioners can discuss with each other matters of practical importance to them and in doing so help their hosts understand what they need to know about ICTs, organizations, and change. Such an arrangement takes advantage of the knowledge and experience of ICT professionals who are "on the front-line of the information technology wars in organizations" (McKeen & Smith, 1997,

p. xi). It provides them with an opportunity to interact with each other, share their insights and experiences, and develop a "wish list" of topics that they would like to see researchers investigate. Each meeting could be focused on a single topic and, guided by their hosts, participants would be encouraged to discuss the ways they have addressed the topic in their organizations. Through conversations with professionals who design and manage ICTs in organizations throughout the economy, SI researchers would be able to learn about the problems and challenges that characterize ICT systems development, the use in different economic sectors, and about the trends that are shaping professionals' jobs. This would be a way for SI researchers to shape their work in ways that would have practical relevance for this group. In return for their participation, IT practitioners would receive research reports about the topics they had selected that would incorporate the discussions with researchers, relevant SI research, and practical strategies that could be used to address the topic in their organizations.

A successful example of this strategy took place at the School of Business at Queen's University in Kingston, Ontario, where two professors ran the "IT Management Forum" during the 1990s. McKeen and Smith (1997, p. 2) periodically brought senior IT managers together to discuss an issue that had been selected by the participants. Forum sessions followed the following format:

- Members jointly select a topic of interest to them.

- They research the topic within their own organization following an outline prepared by the facilitators.

- Each member makes a presentation about how his or her organization is managing the issue.

- Members have an opportunity to critique and discuss each presentation and to agree on best practices.

- Draft copies of each paper are reviewed by members as well as selected IS and user managers in the members' organizations.

- Responses are incorporated into the final papers.

These meetings provided the participants with practical strategies based on relevant research that they could then implement in their organizations. The meetings provided the researchers with a set of research problems based in the practitioners' work that they could investigate. The reports that resulted from this work were made available to the participants who then implemented the relevant findings in their organizations. This relationship allowed research to directly affect practice and worked, in part, because the impetus for the research was provided by the CIOs. Membership in this forum was limited and fee based; annual participation cost $5,000 (CA) and access to the reports was another $1,000 (CA). McKeen and Smith (1997, p. 2) report that the reaction to the reports was largely positive.

6.3.3.5 Providing Continuing Education for ICT Professionals

How can SI research be used to inform continuing education for ICT professionals? What types of workshops and seminars will be of the most value for them?

Another way to reach ICT professionals is through an ongoing series of workshops and distance education experiences. Accepted conventional wisdom among ICT professionals is that the pace of technological change is increasing, and many acknowledge the difficulty of staying current in their specialties. SI researchers can take advantage of this belief by sponsoring workshops based around the theme of using their research to improve organizational practice. This experience should involve both the continual upgrading of technical skills and the ongoing discussion of the importance of the social and organizational context of ICT design and use. According to Fletcher and colleagues (Fletcher, Rollier, Small, & Wildemuth, 1995):

> It must be recognized that it is not possible to teach students all they need to know for a successful career within the confines of a four-year bachelors degree program or a two-year masters. We have long proclaimed our belief in lifelong learning, but we have not adapted our curricula to that belief. Curricula need to be developed with a model of change as its underpinning.

Reflecting on MBA education, Gosling and Mintzberg (2004) argue that "companies and business schools must work together to reinvent management education, rooting it in the context of managers' practical experiences, shared insights, and thoughtful reflection." They believe that a successful learning experience must be based around practice; in fact, they would prefer that management education be "restricted to practicing managers, selected on the basis of performance." The importance of continued education for ICT professionals cannot be overstated. Several years ago, the Tavistock Institute (1998) hosted a workshop on "Planning and Sequencing Successful Organizational Change: An Interactive Workshop Introducing a New Methodology," which was developed for:

> ... senior line managers, human resource managers and in-company change facilitators and managers, working in sectors

undergoing comprehensive change. These include construction, manufacturing, financial services, healthcare or local government.

The purpose of this type of workshop is to provide participants with an experience involving a mix of theoretical insights, empirical research findings, and practical exercises designed to give them the tools and strategies they need to make immediate and observable changes in their organizations. Social Informatics researchers should hold such workshops and involve ICT professionals in the analysis of appropriate cases, and the application of the principles of the workshop to the participants' individual situations. A workshop series for ICT professionals could be sponsored by a professional academic association to which SI researchers belong, such as the Association for Information Systems (AIS) or the American Society for Information Science and Technology (ASIST).

6.3.3.6 Creating Research-Based "ICT Extension Services"

How can "ICT extension services" be used to translate SI research for ICT professionals?

An intriguing strategy is to create "ICT extension services," built on the model of agricultural extension services.[1] Agricultural extension agents work with members of their communities educating them about the latest developments in agricultural research and helping them with problems that arise during the course of agricultural work. Some agents develop outreach programs that involve different types of audiences. For example, The University of Connecticut Cooperative Extension Service's "educational programs and materials cover a wide range of subjects, with emphasis on animal health and agriculture, family/community/4-H Youth development, food/nutrition/food safety, natural resources/land use/environment, and plant agriculture/horticulture/gardening" in New Haven, Connecticut, and eight surrounding counties. The services are provided by extension educators who "are UConn faculty members who extend the University to the community by teaching new skills and providing problem-solving services based on University research. They also keep the University informed about research and informational needs of the community" (University of Connecticut, n.d.).

What extension services share, however, is the commitment to translating research done in the universities with which the researchers are affiliated into terms that are practical and useful to their constituencies. For example, the mission of the Agricultural Extension Service of Nashville and Davidson County (2003), Tennessee, is to " ... extend ... the research and knowledge from the Land Grant Universities (UT and TSU) to the residents of Davidson County." In more specific terms, at the University of Arizona

(2002), agricultural extension work takes place in "classes, seminars, and workshops; public presentations; with newspaper columns and stories; on radio; on the telephone; on the worldwide web; on video and through research-based publications. For 85 years we have delivered educational programs to the people of Arizona." The Department of Agricultural and Extension Education at Michigan State University is "focused on the providing of expertise in the areas of pedagogy and communication, ... and facilitating learning for a diverse group of learners in formal and non-formal settings" (Michigan State University, 2002).

Agricultural extension agents are trained in universities; in its Department of Agricultural Education and Studies, Iowa State University offers an "Agricultural Educators' Professional Development Program" which prepares students for extension work by teaching them about extension history, operations, program planning and development, 4-H programs, instruction and evaluation techniques." They also "have the opportunity to acquire up-to-date technical agriculture knowledge, earn graduate and/or license renewal credit, attain new teaching strategies, develop and receive instructional materials, share curriculum ideas, create links to agribusiness/industry, mentor new teachers, and more" (Agricultural Educator's Professional Development Program, 2003). Training is critical to the success of an extension office; according to an FAO report (Qamar, 1998):

> The positive relationship between training of extension personnel and their performance, both in office and in the field, is well known. To a great extent, the effectiveness of any agricultural extension service is determined by the competence and qualifications of its staff. Studies have shown that the improvement in farmers' knowledge, skills, attitude, efficiency and productivity are positively correlated to the training level and quality of extension staff.

How then can this model be appropriated by SI researchers? An "ICT extension service" based at state universities where SI researchers are on the faculty may provide a very interesting way to bridge the gap between researchers and ICT professionals in small to medium enterprises. Just as farmers are not expected to read the agricultural research, perhaps ICT professionals should not be expected to read academic SI research. The mission of this type of extension service would be to gather the relevant research on ICTs and repackage it for its constituents. Extension agents would be involved in outreach to the ICT professionals in organizations that might not be able to afford the rates charged by large research and consulting firms, but need access to the type of research and services that such firms can provide.

The agents would be, in effect, problem-driven information brokers who would be able to read the research, extract the relevant insights, and develop the practical implications of the work in ways that help ICT professionals improve their practice. They would be able to go into ICT professionals' organizations, provide demonstrations, and give presentations about the SI research relevant to these audiences.

These extension agents would be trained by their universities—perhaps in programs that emphasize both the grasp of ICTs and a deep understanding of information technologies, organizations, and social change. Two examples of programs that could be used to train ICT extension agents are the Masters in Information Science at Indiana University (www.slis.indiana.edu/ Degrees/mis/misindex.html) and the Masters in Information Resource Management at Syracuse University (http://istweb.syr.edu/design/ academic/degrees/grad/irm/index.html). The public universities that would host these extension services should lobby to have a line created for the service in the state budget. The argument can be made that providing this type of service to ICT professionals in small to medium enterprises is as important to the overall health of the state's economy as is support of the state's farmers. Funding this type of initiative at levels equivalent to those provided for agricultural extension programs would be an investment that would pay off handsomely for the states that develop these services.

6.3.3.7 Managing Competition with Research and Consulting Firms

How can SI researchers differentiate their work from that produced by research and consulting firms? Under what conditions is it sensible to cooperate with these firms?

One way to manage the competition with research and consulting firms for the attention of ICT professionals is for SI researchers to distinguish the ways in which their work differs from the work of these firms. They can emphasize the ways in which their research provides a different type of value for practitioners. The types of findings and insights that constitute the best of SI research are qualitatively different from those produced by these firms and can have a different impact on ICT professionals' practice. An important difference is that this research is embedded in a theoretical framework that allows its findings and insights to be generalized so that, according to the CSTB (1998):

> The value of social science research comes not from tracking the frequency of use of the latest technologies but rather from helping to develop common social and economic principles that can be applied to new circumstances. Those designing or relying

on technology and those making policy decisions about the use of technology without reference to systematic theories of human behavior or economics will likely find themselves approaching each new issue in ignorance.

ICT professionals can gain insights from SI research that is not typically found in the reports prepared by research and consultant firms. This type of research is grounded in a broader conceptual framework than a typical consultant's report and has a level of generalization that is not a goal of the consultant. For example, research about how people in organizations are actually using ICTs in their work, the types of problems that they have with existing and new information systems, and the ways in which their organizations are affected by the implementation of new systems can help ICT professionals broaden their understanding of the interrelationships among ICTs and their social and organizational contexts. This, in turn, can enrich the design and development processes in ways that may allow ICT professionals to improve the work lives of the people who use these new systems. In contrast to the type of research produced by research and consultant firms, this implies that SI researchers (CSTB, 1998):

> Do not simply count things. When they count (to answer how much, how many, how frequently), a theoretical context, involving systematic theories of human behavior, motivates the counts ... [however,] the goal is not to count, but rather to understand how the frequency of one kind of behavior affects the dynamics of a social institution. Such understanding enables more reliable forecasts and more trustworthy inferences about causal relationships.

An advantage that academic SI researchers have in comparison to consultants is that their work provides a "socially rich" account of ICTs, organizations, and social change. This is in contrast to consultants' reports, which are usually "socially thin" (Lamb & Kling, 2003). Socially rich accounts conceptualize organizations as environments within which people are enmeshed in a web of dynamic and complex social relationships characterized by subtle interdependencies, periodic conflicts, and multiple interpretations (Lamb & Kling, 2003). This type of approach emphasizes the importance of such social elements as incentive and reward structures, differential social statuses, power relationships, and mutual interdependencies as factors in understanding the uses of ICTs in organizational settings. These elements are glossed over in the socially thin accounts produced by consultants, which can lead to a conception of work practices that does not pay adequate attention to work practices, work-flow, and routines.

A second way to manage the competition with research and consulting firms is to work with them when possible. SI researchers can make their work available to these firms, which are likely to have people on staff whose job is to scan and work with the relevant academic literature. Once they locate research that appears to be useful to the firm, they repackage it and disseminate it to the consultants who can use it. Researchers can also send their work to large organizations, which are also likely to have people who follow academic research. In effect, the work of translation from academic to practitioner-oriented language can be outsourced. Research and consulting firms can be used to bridge the gap between SI researchers and ICT professionals.

The implementation of the first two of these six strategies will take time, given that sustained research efforts will be required to develop an adequate understanding of the information environments of ICT professionals. SI researchers will have to consider carefully the costs and benefits of refocusing research agendas on the problems and issues of importance to this audience. Exploring the idea of ICT extension services will also take some time. The remaining strategies can be enacted immediately. Researchers can begin writing for and speaking to the ICT professional audience now. Workshops and seminars can be conducted if sponsors can be found. Forums can be started that bring ICT professionals into academic settings, and SI researchers can begin to manage the competition with research and consulting firms. Sometimes this will involve distinguishing their work from that done by consultants and other times by building contacts with these firms and sharing their research. The next section examines strategies for reaching a second audience, members of academic and research communities outside of SI.

6.4 Communicating to Academic and Research Communities

6.4.1 Audience

Social Informatics research should also be communicated to research communities across a range of academic disciplines that share a focus on the social contexts of computing, including information science, computer science, information systems, sociology, anthropology, instructional systems technology, management information systems, journalism, and communications. Regular and pervasive communication of research is an important component in raising awareness among academic researchers about the themes, insights, and findings of SI. This is a sensible move because SI's core concepts, frameworks, and findings cut across many of these disciplines and increasing the familiarity of academics with this work can improve their research and teaching. This effort is necessary to legitimize SI among the wider academic community and bring its concerns to the attention of potential funders of this research. It can also help to grow the SI community, as researchers and teachers outside of SI begin to learn more about the field and to explicitly identify themselves as belonging to the community. Many ICT-oriented disciplines already have researchers who are investigating questions and problems that characterize SI. Strategies are needed to bring their work to the attention of their colleagues and to cross-pollinate their disciplines with relevant research from other disciplines.

6.4.2 Challenges of Communicating to Academic and Research Communities

One main challenge to be faced in communicating SI research to a range of academic and research communities is crossing the boundaries among disciplines. Scholarly communication typically takes place within well-defined communities of discourse, each of which has its traditions, literature, research issues and controversies, theoretical and methodological approaches, and conceptions of criteria for adequate research. The gate-keepers of these communities, journal editors and referees, understand these norms well and use them effectively to maintain the standards of quality in their disciplines. According to Cronin (1999), "disciplines differ significantly in terms of their sociocognitive structures, degrees of paradigmatic consensus, funding mechanisms, collaborative intensity, and institutionalized quality assurance mechanisms." One consequence of this situation is that researchers within a community of discourse will draw upon the resources of their community in their work more heavily than they will upon others. Citation studies bear this out, indicating that cross-disciplinary patterns of citations in research articles are less common than patterns of citation that remain within the discipline of the author's primary affiliation (Miyamoto, Midorikawa, & Nakayama, 1990, p. 81). In practical terms, this means that a researcher in library and information science will have a difficult time placing an article in a management journal because the work will not "look like" what editors and referees are used to reading.

To cross academic communities of discourse requires boundary spanning activities. At individual levels, boundary spanners are researchers and theoreticians whose work is respected and used by scholars in many disciplines and whose professional networks range across the academic landscape. Their work overlaps with that of scholars outside their discipline of primary affiliation. These individuals can publish and present their work in these disciplines and are valuable actors in scholarly communication because they can spread ideas across disciplinary boundaries. They can stimulate the types of research and theorizing that create links and channels of communication across disciplines. However, such individuals are rare. Institutional efforts to encourage boundary spanning are also valuable and increasingly important because they offer opportunities for researchers from different disciplines to interact and learn from each other. For example, the National Science Foundation has sponsored a series of workshops that have had the explicit goal of bringing together researchers from a range of disciplines who share interests in common topics such as digital libraries, electronic commerce, and Social Informatics. How can boundary spanning activities be routinized?

6.4.3 Strategies for Improving Communication with Other Academic and Research Communities

What can be done to improve the communication among SI researchers and scholars and researchers outside of this community? How can the profile of SI research be raised across the wide range of disciplines in which SI themes are present? How can scholars and researchers who are currently outside of SI be encouraged to join and participate in the community?

In this section, five strategies are proposed that can be used to communicate SI research to academic and research audiences. The first is to raise the profile of the research at academic conferences. The second involves expanding the publishing options for SI researchers, providing them with opportunities to bring their work to the attention of scholars outside of their primary disciplines. The third strategy takes advantage of the well-connected digital information environment within which academics routinely interact to provide easy access to research and other information about SI activities. The fourth involves a series of research initiatives, and the fifth is based on developing greater institutional support for SI research at colleges and universities.

In summary, the five strategies are:

1. Raise the profile of SI research at academic conferences.

2. Expand the publishing options for SI researchers.

3. Take advantage of widespread and easy access to network-based digital information.

4. Engage in a series of research initiatives that will produce useful outcomes that gain attention for SI research.

    5. Develop greater support for SI research at colleges and
universities.

6.4.3.1 Raising the Profile of Social Informatics Research

*What can be done to bring SI research to the attention of researchers and
scholars outside of the field?*

Three initiatives are proposed to raise the profile of SI research at academic conferences. These include the creation of SI special interest groups
(SIGs), the organization of research tracks and panel discussions of SI
research and issues, and the development of an entire conference devoted to
SI research.

Established SI scholars can set up SIGs in professional organizations,
drawing interested members together, providing them with regular opportunities to interact. For instance, organizations such as the Association of
Computing Machinery (ACM) or the Association for Information Systems
(AIS) can be approached about starting an SI SIG. There may be a way to
combine SIGs from a variety of related groups to build a large enough population to cosponsor conferences. Umbrella or confederation groups might be
another way to pull the field together. Group leaders could plan activities or
tracks at existing conferences that might attract more people than small specialty conferences.

Second, members of various professional organizations such as the ACM,
ASIST, and the International Conference on Information Systems (ICIS) can
begin to set up research tracks at their regional, national, and international
conferences. Such tracks provide a venue for researchers who wish to identify themselves with SI and can thus bring SI research to the attention of a
wider academic audience. In recent years there have been SI tracks and panels at a number of national and international conferences. For example, there
was an SI track at Association for Information Systems America's Conference
on Information Systems in 1997 and 1998. Here is a partial listing of recent
conference activity:

> Eschenfelder, K. R. (Organizer), Hara, N., Lamb, R., Rosenbaum,
> H., Sawyer, S. (October, 2004). A Social Informatics Workshop
> for Library and Information Science Research. In *Proceedings of
> the 67th Annual Meeting of the American Society for Information
> Science*, 41. Medford, NJ: Information Today, Inc.

> Rosenbaum, H. (Organizer and moderator), Davenport, E.,
> Eschenfelder, K., Lamb, R., Murphy, L., and Sawyer, S. (August,
> 2004). Panel: Social informatics and information systems

research: Where are we going? In Romano, N. (Ed.), *Proceedings of the 10th America's Conference on Information Systems.*

Kling, R., and Lamb, R. (1998). Advances in Social Informatics and Information Systems. Association for Information Systems 1998 America's Conference (20 papers). http://www.isworld.org/ais.ac.98/proceedings/si.htm

Rosenbaum, H. (Organizer), Davenport, E., and Sawyer, S. (1998). Advances in Social and Organizational Informatics: Implications for Information Science Panel session at National Meetings of the American Society for Information Science, Pittsburgh, PA.

The First Kyoto Meeting on Social Interaction and Communityware (June 8–10, 1998, Shiran Kaikian, Kyoto, Japan). http://www.lab7.kuis.kyoto-u.ac.jp/km

Kling, R., Rosenbaum, H., and Travica, B. (1997) Mini-Track: Advances in Social Informatics for Information Systems. Association for Information Systems 1997 America's Conference (6 papers). http://hsb.baylor.edu/ramsower/ais.ac.97/program.html

IRIS 20: Social informatics. Hank Fjordhotel, Norway, August 9–12, 1997. http://www.ifi.uio.no/iris20/conference.html

Should there develop a critical mass of scholars and researchers interested in SI issues, a third initiative is to host a series of high level conferences inviting a group of people acquainted with the relevant literature, who then engage in common research and theoretical discourse, to evaluate the current state of the field. An annual conference would have the effect of moving SI away from an identification with one institution or individual and could include a support network for doctoral students, as well as opportunities to bring people and ideas together from the various disciplines concerned with the social and organizational impacts of ICTs on work and social life.

The conference would be interdisciplinary, relatively small (about 100 people) and would have a research focus. It could be paired with a larger and a more established conference in which there would be some overlap. There could be pre-conference workshops on pedagogy or teaching, and a doctoral consortium.

6.4.3.2 Increasing Publishing Options for
Social Informatics Research

Which publishing outlets can be used to disseminate SI research more widely within the academic research community?

The second strategy is to increase the options for publishing SI research by targeting appropriate journals, writing textbooks, and editing collections of exemplary SI research.

There should be a concerted effort to identify and target relevant journals that have significant impact among their readers and that would be interested in publishing SI research. A list of such journals could be compiled by individual researchers and submitted to a central location for compilation, such as the Rob Kling Center for Social Informatics Web site (http://www.slis.indiana.edu/CSI). As a way of sensitizing editors, referees, and readers to SI issues, researchers should edit special issues of prominent journals in different disciplines. For example, a "Special Issue on Social Informatics" in the *Journal of the American Society for Information Science* presented seven research papers that outlined some current issues and findings of SI for an information science audience (Kling, Rosenbaum, & Hert, 1998). Where possible, articles about SI can be placed in journals that will bring the discipline to the attention of a wider academic audience. Kling (1999a) wrote an article on SI, "What Is Social Informatics and Why Does It Matter?" for *D-lib Magazine*, a publication read by members of the digital libraries research community.

There also should be an effort to develop a corpus of literature around SI's core. To reach the next generation of academic researchers and ICT professionals, there is a need for edited volumes of SI research and cases for classroom use. Researchers should consider writing textbooks that present the SI core in ways that assist teachers; this is one good strategy to help define a field. This strategy is developed in detail in Chapter 5. Compilations of SI papers from a variety of conferences would be a useful reference source for researchers and teachers. Participants at the NSF workshop on "Advances in Social Informatics" suggested the creation of a scholarly "Social Informatics Handbook" with articles that would be commissioned from experts in advance. This volume would be targeted at researchers. It would contain "state of the art" literature reviews, research exemplars, concise statements of major theoretical approaches in Social Informatics, and current bibliographies.

6.4.3.3 Taking Advantage of Easy Access to Networked Digital Information About Social Informatics

How can SI researchers exploit the Web to disseminate their work more widely? What type of model should be used to create a vibrant online Organizational Informatics and Social Informatics community?

The third strategy is a set of computer-mediated initiatives that takes advantage of the digital information environment within which many academic researchers work. These initiatives involve the creation of an online SI community, with a Web site and other forms of digital communication, such as listservs and newsgroups. Such a community could be built along the model used by ISWORLD (http://www.isworld.org/isworld.html), which is supported by professional organizations and educational institutions and is run as a distributed web. ISWORLD is an online community for information systems academics that provides research and teaching resources to its members. Faculty members at different educational institutions bid for the responsibility for organizing and maintaining specific sectors of the web. For example, as of the writing of this chapter in summer 2004, the Qualitative Research sub-web is maintained at the University of Auckland (http://www. qual.auckland.ac.nz), the Teaching and Learning Division is sponsored by the Grand Valley State University (http://www.magal.com/ iswn/teaching), and the Research Task Repository is maintained at the Kelley School of Business at Indiana University (http://kelley.iu.edu/bwheeler/ISWorld/ index.cfm). There is also a set of active online discussion groups and a listserv that is used to disseminate information and announcements of interest to the IS community.

Such a site for SI researchers could take advantage of the digital infrastructure already in place. It should include links to other professional organizations that have well-developed and relevant Web sites, including the IAMCR (mass communication research in Europe), the TPCR (telecommunications policy research conference), OCIS, and IRIS (Scandinavian conference on Social Informatics). The beginnings of an SI Web site can be found at the Rob Kling Center for Social Informatics (CSI) at Indiana University (http://www.slis.indiana.edu/CSI), which "is dedicated to support research into information technology and social change." There is a listing of the Fellows of the Center and the doctoral students who are conducting dissertation research on SI topics. There is a growing collection of working papers, many of which are available in full text with annotated lists of SI workshops, seminars, conferences, and journals. This site can serve as a central hub for information about SI and can be a point of access for research, academic programs, courses, and a wealth of other resources that focus on SI concerns. The RKCSI Web site is a small and growing resource for information about SI, but it needs support to grow into a technology that can support a community.

There should be a directory of SI faculty that includes names, mail and e-mail addresses, and phone numbers so people can learn who is involved in the field and how to contact them. Initial seed money could be used to produce the first directory. To follow the ISWORLD model, it will be necessary to create a distributed network of SI Web sites at different locations, each with the mission of developing some subsection of the site. Workshop participants suggested creating and maintaining a set of data and other information about SI. The data would be in digital form and in the public domain so that interested academics could begin to learn about the type of work that characterizes the field.

Three mailing lists have been set up for people interested in SI. Once these lists gather a critical mass of subscribers, another channel of communication to open is in community building across academic disciplines. The first, SI-ANNOUNCE-L@listserv.indiana.edu, is a one-way mailing list for announcements of conferences, workshops, calls for papers, and the like that are related to SI. The second, SI-TEACH-L@listserv.indiana.edu is a discussion list for issues concerning SI teaching, including teaching methods, curricula, and teaching materials. The SI-RESEARCH-L@listserv.indiana.edu mailing list is intended to be used as a forum for communication among people who are interested in the field of SI. Its focus is on the issues of developing the field of SI with an emphasis on the development and methodology of research and research issues in the field.

6.4.3.4 Research Initiatives to Raise the Profile of Social Informatics

What types of large-scale collaborative research initiatives can be undertaken to increase the awareness of SI research among the wider academic and research-funding community?

The fourth strategy involves a series of research initiatives designed to raise the profile of SI research in the academic community. A team of SI researchers should convince a funding organization to sponsor a national or international survey on ICTs in organizational settings and play a prominent role in directing the research. The results should be widely publicized and made available in a variety of formats. Subsequently, a professional SI organization could sponsor the survey on an annual or biennial basis, so that it could become a reliable "science and technology" indicator. This would be a valuable source of information for policy makers, researchers, and teachers in a wide range of disciplines. Small subcommittees or teams can be developed to work on different parts of the survey.

The survey could cover the social impact of technology and technology dissemination and be done at the individual/household/business/organization level—perhaps at a different level each year. There could be different questions for the different levels. The survey could cover infrastructure issues, households, spending patterns, and so forth. There could be questions on opinions about technology, spending and usage patterns, and quantitative measures of infrastructure. It would be nonpartisan and sponsored by universities rather than corporations. A workshop could be held to construct the survey so that many researchers could contribute questions from their research. The survey results would encourage policy makers to recognize the field and to fund further SI research. This type of project would be useful to researchers and would help promote the field to the National Science Foundation in the United States and equivalent funding organizations in other countries.

6.4.3.5 Increasing Institutional Support for Social Informatics Research

How can SI researchers encourage cross-institutional collaboration? What can they do to encourage their institutions to increase funding for their work?

The fifth strategy is based on developing greater institutional support for SI research at colleges and universities. SI researchers should encourage their institutions to sponsor faculty exchanges, allowing interaction to occur across organizational and perhaps disciplinary boundaries. Such exchanges could be beneficial to universities because they provide a means by which institutions can fill temporary vacancies with visiting scholars and researchers on sabbatical. This would save the universities money and help promote SI research.

If implemented, these five strategies can begin to routinize the boundary spanning activities that are necessary for SI researchers to be able to communicate easily with other academic and research communities. They can also make it easier for researchers and teachers not affiliated with SI to learn about the research, programs, courses, and other resources that constitute the corpus of this field. The first three strategies could be done with relatively low cost in terms of time and money. SI researchers are already raising the profile of their research at academic conferences by sponsoring panels and research tracks. There has been a special issue of a major professional journal focusing on SI, and articles have appeared in a variety of journals, expanding the publishing options for SI researchers. The Rob Kling Center for Social Informatics Web site is continually under development and the SI listservs are online, taking advantage of widespread and easy access to network-based digital information. The last two strategies are more difficult and will take time, given the efforts involved in coordinating a large scale research initiative, such as an ICT indicator survey, that will produce useful outcomes that gain attention for SI research. In times of tight budgets, developing greater support for SI research at colleges and universities is always a struggle, but is part of an ongoing effort in which researchers are always engaged.

6.5 Raising the Visibility of Social Informatics

This chapter has discussed the communication of SI research to professional and academic research communities. It has argued that SI researchers should increase their efforts to communicate their research, insights, and findings to ICT professionals because the knowledge that they can impart can have real and immediate value to these practitioners. As the costs of ICTs and information systems increase, ICT professionals will come under increasing pressure to ensure that the systems they design and implement maximize the organization's return on investment. If these practitioners continue to hold to the magic bullet theory described by Markus and Benjamin (1997), the failure rate of systems implementations is not expected to decline. The insights and findings of SI research are a corrective to this view, but if they cannot be persuasively communicated to ICT professionals they can have no influence on their professional practice.

We have suggested here that in order to reach the audience, two main challenges need to be surmounted. The first is perceptual, and involves a belief among ICT professionals that SI research is not relevant to their work. The second is competitive, because SI researchers are not the only group that is trying to communicate with ICT professionals. Research and consulting companies pose a threat to the ability of SI researchers to gain the attention of ICT professionals. A series of strategies were suggested to help SI researchers communicate more effectively with this audience:

- Develop a current empirical assessment of what we know about the worlds of ICT professionals.

- Conduct and communicate research that is useful for these professionals given what we know about their typical problems, concerns, and information behaviors.

- Focus on speaking to and writing for ICT professionals.

- Hold regular forums that bring together academics and ICT professionals.

- Influence current professional practice through workshops, seminars, and life-long learning.

- Create "ICT extension services."

To manage competition with research and consulting firms, SI researchers can:

- Distinguish their work from that done by the firms, emphasizing the value that can be added to practice from academic work.

- Explore ways to cooperate with these firms by sharing research.

The long-term goal of this effort is to persuade ICT professionals that they can rely on SI researchers for high quality research findings and insights that will be useful in their work. The professional ICT audience should look routinely to SI researchers for ways to improve their practice and deepen their understanding of the complex interrelationships among ICTs, the people who work with them, and the organizations in which they work.

To reach members of the academic community, the main challenge is to cross the boundaries that separate disciplines. Individual boundary spanners are valuable in this effort; however, they are rare and are typically "found" and not "trained." Five strategies are suggested to improve the communication of SI research to other academic and research audiences:

- Raise the profile of SI research at academic conferences.

- Expand the publishing options for SI researchers.

- Take advantage of widespread and easy access to network-based digital information.

- Engage in a series of research initiatives that will produce useful outcomes that gain attention for SI research.

- Develop greater support for SI research at colleges and universities.

These strategies will improve the ability of SI researchers to make their work known to others outside of their own communities. They will also provide many opportunities for researchers and teachers outside of the community to learn about SI, encouraging them to incorporate SI themes, theories, concepts, findings, and insights into their research and teaching. These strategies may also help to grow the community by encouraging more researchers and teachers to explicitly identify themselves as being interested and involved in SI.

This work will not be easy and it will take time and concerted effort by SI supporters. It will be critically important in the process of building the field and raising its profile among two of the communities that must be reached: ICT professionals, and academic researchers and teachers.

Endnote

1. In the United States, agricultural extension services tend to be affiliated with land grant universities. They operate as quasi-independent agencies with an explicit outreach function.

CHAPTER 7

Conclusions

Throughout this book we have outlined and illustrated both the conceptual underpinnings and common findings of Social Informatics research. We have made the case that Social Informatics research: 1) is a growing body of literature, 2) spans existing academic disciplines, and 3) is that which examines the design, implementation, and use of information and communication technologies (ICTs) in social and organizational settings and the roles of these in social and organizational change. In doing this, we have summarized representative works from the published literature that contribute to Social Informatics findings and concepts. We have shown how this research base is critical for informing our conceptualizations of how ICTs are designed, developed, and used; the ways in which ICTs are conceived in the public policy debates about such use, and current professional education and practice focused on preparing future ICT-oriented professionals.

In doing this we have showcased how the growing, dispersed, and increasingly relevant body of Social Informatics research provides the knowledgeable professional with current analytic skills, empirically grounded concepts, and common findings. These skills, concepts, and findings are needed to properly design and configure ICT-based systems so that they are actually workable for the people for whom they are intended.

We have further argued that the hallmarks of Social Informatics research—findings built on careful, contextually situated, and empirically grounded analysis—contrast, often dramatically, with high-spirited but largely a-priori promotions of new ICTs by vendors, pundits, and uncritical analysts. We have used evidence from the contemporary academic literature to illustrate that such optimism often leads to the design and implementation of ICTs that may occasionally work well for people and may occasionally be valuable, but are too often abandoned or unusable, thus incurring needless waste and inspiring misplaced hopes in the capabilities of ICTs to bring about positive changes in the workplace or home.

As we noted at the outset, findings drawn from careful empirical research may not be as catchy as a popular pundit's enthusiasms. However, the cumulative findings drawn from Social Informatics research provide a reservoir of learning that helps to frame and analyze the realistic possibilities and potential effects of new ICTs—often in contrast to popular conceptualizations that characterize these effects as innovative and revolutionary.

Beyond this quick summary, we have other points to make in this final chapter. First, we review the cumulative findings from Social Informatics research. Second, we emphasize particular findings that are relevant to four

145

groups of people: designers and developers, policy analysts, educators, and scholars of ICTs. Third, we point out a number of contemporary efforts to theorize about ICTs that are grounded in a Social Informatics perspective. In doing this we note that this is a representative, not exhaustive, review of the types of research problems that are increasingly the focus of Social Informatics scholars' attention. Fourth, we argue that Social Informatics researchers have an increased professional responsibility to disseminate their work to academic, ICT policy-oriented, and professional practitioner audiences. We end the chapter and the book with a short section in which we note that Social Informatics represents significant scientific progress in our understanding of ICTs, the people who design and use them, and their contexts of design and use.

7.1 Summary of Findings, Concepts, and Issues

The value of a Social Informatics perspective is illustrated by the ideas developed in this book, the most important of which are summarized briefly below.

7.1.1 ICTs Are Socially Shaped

From a Social Informatics perspective, ICTs are highly intertwined in socio-technical networks in which software features, hardware elements, roles, people, social and organizational culture and structure, and norms and rules of use are intimately interconnected (see Table 2.1 in Chapter 2, and Kling & Scacchi, 1982). This conceptualization focuses on computing as a complex activity that is tied together in a web of socio-technical practices and resources (e.g., Avison & Wood-Harper, 2003; Avgerou, 2002; Barley, 1986; Dutton, 1999; Suchman, 1987). For example, the types of social incentives that are present affect the use of new electronic media in an organization; these can range from an organizational culture that encourages and rewards ICT use to one that dampens enthusiasm for the new technology. To argue that ICTs are socially shaped means that to study ICTs, researchers must attend carefully to the people who design, implement, and use them and to the social and organizational contexts of their design and use.

7.1.2 Problem-Oriented Nature of Social Informatics Research

Social Informatics research is problem oriented. Just as the human–computer interaction (HCI) literature reflects the problematic relationships between individuals and computers, and the computer-supported cooperative work (CSCW) literature reflects the problematic relationships between people who work in groups and the computer systems they use (e.g., Ackerman, 2000), the Social Informatics literature reflects the problems that arise from the bi-directional relationship between social and organizational contexts and ICT design, implementation, and use. Social Informatics research spans levels of analysis, often by making explicit links between particular levels of social analysis and the larger social milieu in which computing takes place. In this way, Social Informatics is similar to other areas of study that are defined by a problem such as gerontology, software engineering, and urban studies. Social Informatics work is done in a wide range of settings such as communities, complex organizations, physical and digital libraries, hospitals, geographical areas, and virtual spaces. This contrasts with disciplines such as public administration, business/management, health, and education that are focused on specific domains or aspects of a domain. Moreover, Social Informatics research is not defined by analytic methods or study domains.

7.1.3 People Are Social Actors

In Social Informatics research, people are explicitly characterized as "social actors" (e.g., Lamb & Kling, 2003, and Table 3.1 in Chapter 3) and not simply as "users." That is, people are seen as acting routinely and competently within the social and organizational settings through which they pass as they carry out their work and social activities. As social actors they have some, but not complete, individual agency and live and act within a shifting set of social structures and norms that, while fluid, shape behavior in both subtle and direct ways. This social actor perspective is often represented through the use of social and socio-technical theories (Horton, Davenport, & Wood-Harper, 2005; Sawyer & Eschenfelder, 2002). For this reason, Social Informatics researchers do not adopt assumptions of technological determinism in their work. Through their actions, people participate in both social stability (maintaining the status quo) and social change.

7.1.4 ICT Use Is Situated and Contextually Dependent

The mutual interdependencies among ICTs and their social and organizational contexts frames much of Social Informatics research (see Table 5.1 in Chapter 5). By emphasizing the theoretical and methodological centrality of social and organizational contexts we argue that Social Informatics researchers typically employ a holistic perspective. In their work, they attend to a range of specific characteristics that help to define a level of analysis, act as forces on the various levels of analysis, and provide the backdrop and perspective from which an understanding of the problem of interest can be developed. The exact nature of the social and organizational context is intimately related to the problem of interest: Studies focused on regional implementation patterns of GIS will depict the context differently than will studies that explore the distributed work of air traffic controllers in combat situations. This suggests that the characterization of, and factors of interest within, contexts will vary as the research focus shifts; the researcher must set out the levels of analysis and factors through either *a priori* depiction or *post hoc* description. All Social Informatics research assumes the importance of social and organizational contexts as a key to understanding ICT design, implementation, and use.

One way to place these four basic ideas of Social Informatics into perspective is to acknowledge that they stand in sharp contrast to the more common and too simple, tool-based models of ICT described in Chapter 1. We have noted that tool-based models of ICT are always incomplete and very often inaccurate. As we develop more elaborate ICTs and begin to use them in almost every sphere of social and organizational life, the shortcomings inherent to the tool-based view of ICT make this an increasingly inadequate, if not dangerous, way to approach computerization. Simply put, we face fresh theoretical challenges, challenges that the tool-based view of ICT is not able to address. This is particularly cogent even if the tool-based model were to have been useful at one point, some time ago. In this book, we have argued that Social Informatics offers an alternative to tool-based views of ICT. We further noted that Social Informatics is, increasingly, a more viable—if not indispensable—analytical foundation that will drive the research forward improving our understanding of ICTs and the interdependencies among the ICT artifact—the people who design, implement, and use them, and the contexts of their uses.

7.2 Specific Relevance to Particular Domains of Interest

In Chapters 3, 4, and 5, we highlighted Social Informatics findings and examples with specific relevance to those involved in designing, developing, and implementing ICTs (Chapter 3), those engaged in information and ICT policy analysis and policy making (Chapter 4), and those involved in ICT education (Chapter 5).

7.2.1 Social Informatics Relative to Designing, Developing, and Implementing ICTs

In Chapter 3, we developed the concept of social design. We contrasted social design with the more common designer-focused approaches (see Table 3.1 in Chapter 3). In doing this, we explored the ways in which design work is characterized, the goals of the ICT design effort, its underlying design assumptions, and the technological choices made during and after the ICT design process. The social design of ICTs is driven by the principles that arise from Social Informatics literature, among which is the insight that design must recognize and accommodate the configurable nature of ICTs. This approach makes explicit the goal that an ICT-based system must be designed for heterogeneous groups of users and their diverse uses. Further, we noted that, because all information systems involve agency, social design principles are intended to make agency more transparent for designers and allow for instances of multiple agency. Finally, Social Informatics research makes clear that design continues into use, suggesting to us that design approaches must more explicitly attend to post-implementation changes in ICT-based systems (Eschenfelder, 2004; Faraj, Kwon, & Watts, 2004; Heeks, 2002; Introna & Nissenbaum, 2000; Kling & Iacono, 1984; Markus, 1983).

7.2.2 Social Informatics Relative to Information and ICT Policy Making

In Chapter 4, we explored the utility of Social Informatics for policy analysts, focusing on the ways in which the concepts and findings of Social Informatics research can help in their work. The main theme of this chapter is that analysts focus on the policy implications of complex ICTs, such as the digital divide or the protection of digital intellectual property. In doing this, they face increasing uncertainty, especially when ICTs become deeply embedded in routine social and work practices. By developing awareness and an understanding of policy-relevant Social Informatics research, ICT policy analysts will be more easily able to disentangle the issues and alternatives raised by new computing-based technologies and ascertain the impacts of policy alternatives on social and organizational life.

Social Informatics research can provide policy analysts with a set of reliable and tested conceptual frameworks and empirically based findings that can be used to isolate, organize, and understand the range of social, cultural, and organizational forces that surround the design, implementation, use, and regulation of ICTs in the social world. Armed with this knowledge, policy analysts can provide policy makers with realistic analyses of the social consequences of different decisions about the regulation of ICTs.

7.2.3 Social Informatics in ICT-Oriented Formal Education

In Chapter 5, we made three points. First, we highlighted that academic, professional, and public-sector institutions predict that there is a need for more properly educated ICT professionals, and that this education must include findings and analytic techniques of Social Informatics. Second, we noted that, although this material is increasingly being taught, coverage is still spotty and Social Informatics literature often appears only in elective courses. Finally, anticipating that providing guidance to help ameliorate this current shortcoming, we laid out seven Social Informatics principles that should be taught throughout any ICT-oriented curriculum. First, students need analytic skills to help them understand how the context of ICT use directly affects the meanings and roles of the designers and users of these technologies. Second, there is a need to adopt a socio-technical perspective to help explore the values embedded in ICTs (incorporating a fundamental assumption of Social Informatics that ICTs are not value neutral). We made the case that it is often difficult for students to understand issues or values and their relationships to technologies when using traditional tool-based models of ICT. Third, ICT use leads to multiple, and often paradoxical, effects and so demands analytic techniques that help to explore alternative scenarios of use and mitigate unintended consequences.

Fourth, the design and use of ICTs have moral and ethical dimensions and these in turn, have social consequences. In short, when ICTs are implemented and used in organizational and other settings, there will typically be winners and losers; some will benefit from the implementation of ICTs, others will not. To grasp this, students must have the skills to conduct ethical and critical analyses of ICT design, implementation, and use. Fifth, ICTs are configurable: Systems are actually collections of distinct components and these components can be arranged and used in different ways, leading to different outcomes, even if implemented in similar types of social or organizational settings. This requires that students have the requisite conceptual understanding of computing and a set of work, organizational, and social analytic skills to aid in design. Sixth, students should understand that ICTs follow trajectories and these trajectories often favor the status quo, suggesting that historical and market analytic skills are important. Finally, students should learn analytic techniques that allow them to guide ICT-based systems as they co-evolve with their users through design, development, and use.

7.3 Moving from Collecting Findings to Theorizing About ICTs

Social Informatics researchers are increasingly interested in developing theories and models that elaborate on the processes and effects of the social design, development, and implementation of ICT-based systems. These efforts build on more than thirty years' work developing reliable knowledge about ICTs and social and organizational change. Further, this attention to theorizing arises from systematic empirical research drawn from a range of literatures and academic disciplines.

7.3.1 Institutional Nature of ICTs

Social Informatics research explicitly engages the larger context in which ICTs exist. These ICTs may be set in formal organizations, workgroups and teams, informal communities, the home, educational settings, online milieus, and larger social settings. Increasingly, the social and organizational contexts of ICTs are being represented as institutional in nature. Institutions are enduring social structures, and are depicted in a range of rich and diverse instances (Scott, 2001). This institutional depiction of social and organizational contexts spans cognitive, normative, and regulative elements, varies by levels of formalization, exists in multiple and overlapping arrangements, and reflects the insight that social structures can be both stable and mutable (Agre, 2000; Avgerou, 2002; Butler, 2003; Fountain, 2001).

Social Informatics research typically shows people participating within and across many institutions (e.g., people are professionals, working in formal organizations, on work teams in particular cultural settings, and may routinely cross organizational boundaries in their work). Further, the findings indicate that these multiple participations influence ICT use and lead to complex and multi-faceted meanings of use.

An institutional perspective on the social and organizational contexts of ICTs helps to explain the complex interactions between institutions and individuals. Lamb and Kling (2003) call these interactions "social action" and argue that people using computers are better understood as "social actors" rather than as "users." As social actors, their actions are typically shaped in ways that lead them to perform routine and legitimate actions and interactions that are defined and regulated by institutional structures; typically, these interactions occur within socially structured arrangements in organizations or other constrained settings of our social lives. This suggests that social actors have a bounded range of choices and that their choices are further influenced by both historical commitments and contemporary forces. However, this approach is by no means deterministic; social actors can make use of ICTs in ways not predicted or intended by designers that can result in social and organizational change.

7.3.2 Conceptualizing Computing from a Social Informatics Perspective

Social Informatics researchers continue to do rigorous empirical research, adding to the rich and growing body of findings. However, a steadily increasing number of those doing work in Social Informatics are now moving from generating and collecting common findings to theorizing and developing analytic approaches that reflect the conceptual elements and empirical findings that we have discussed in this book. For example, Avgerou (2002) embodies this effort in her book, with chapters on the institutional nature of ICTs, their socio-technical nature, and an extended discussion of the ways in which both people and their social and organizational contexts can be rigorously conceptualized. As is the Social Informatics tradition, these chapters focus on theorizing a Social Informatics perspective and are supported by extended case studies. Avgerou's work is but one example, and a more complete review of the theorizing on Social Informatics includes works from a range of scholars, many of whom are cited throughout this book and listed in Appendix A.

7.4 Social Informatics as a Professional Obligation

In Chapter 6, we argued that Social Informatics researchers have a professional obligation to communicate the results of their research and theorizing to their academic peers and to practitioners. The work should have relevance to both professional and research communities. There is a pressing need to communicate the findings of Social Informatics research because of evidence indicating that corporate investment in new ICTs continues unabated, that new ICTs continue to be under-used, that their use often leads to unintended (and even injurious) consequences, and that those ICT-based systems that are integrated into work practices in organizations routinely fall short of expectations.

We argued that two audiences could benefit immediately and immensely by engaging more directly Social Informatics research: professionals involved in the design, implementation, and management of ICTs and researchers and educators in a range of academic disciplines who are concerned with ICTs, people, and work (and social) life.

Reaching the first audience is important because Social Informatics research can improve professional practice. The more professionals understand how ICTs and people interact in organizational settings, the better prepared they will be to manage both. The more designers understand about the complex social and organizational contexts in which the ICTs they are designing will be implemented and used, the better and more usable the systems will be that they will develop.

The main challenge in reaching this audience is to convince them that the research is relevant, timely, and valuable to their immediate work contexts.

Six strategies were suggested to reach this audience. The first is to focus research on the work worlds of ICT professionals. The second is to analyze their typical problems, challenges, and concerns to arrive at concrete suggestions for improving practice. The third is to communicate regularly with ICT professionals in meetings and fora; this requires developing an incentive structure in the academy that rewards such activities. The fourth is to hold regular meetings to bring academic researchers together with ICT professionals. The fifth is to influence current practice through workshops, seminars, continuing education credits, and life-long learning. Finally, Social Informatics researchers should help to develop ICT "extension services," or regional pockets of expertise upon which professionals can call.

Reaching the second audience, academic peers in different disciplines, is also important because Social Informatics cuts across so many disciplines. Researchers interested in Social Informatics problems should be aware of each other's work if for no other reason than to avoid reinventing the wheel or following someone down a blind research alley. Cross-disciplinary communication will also raise the profile of Social Informatics in the academy and bring it to the attention of funders of social science research into technologies and society. One main challenge here is how to routinize boundary-spanning activities. Five strategies are offered to accomplish this. The first is to present research that is explicitly labeled as Social Informatics at academic conferences, perhaps through panels and workshops. The second is to expand publishing opportunities for Social Informatics researchers, perhaps through special issues of prominent journals. The third is to make use of the Web to provide easy access to research and other information about Social Informatics activities. The fourth involves a series of research initiatives such as a multinational consortium to conduct a global survey of ICT use in organizational settings. The fifth is based on developing greater institutional support for Social Informatics research at colleges and universities by creating centers for Social Informatics, as has been done at Indiana University in the U.S. and Napier University in Scotland.

7.5 Taking Social Informatics Seriously

We have laid out the case for Social Informatics moving from being a useful but alternative approach to being one that is the core knowledge basis for understanding computing. The rationale for taking social and organizational informatics seriously is that by doing this we will improve the professional practices of designing, developing, deploying, and using ICT and ICT-based systems. Systems designs informed by Social Informatics will be more useful, ICT policy will be more effective, and future researchers will engage more substantial questions and provide more useful insights than those who ignore the interconnections among the social and technical aspects of computing.

Through this monograph we have introduced you to the systematic, rigorous, and empirically based research that has focused on these and other computerization issues. Further, in organizing the collected findings conceptually in the main chapters of the book, we provide you with a means to draw on the large and growing body of research that addresses the interrelationships among ICTs, the people who design, implement, and use them, and the varied contexts in which they are used.

Through this book we have moved beyond the pleasant simplicities that pervade the public discourse on how ICTs change organizational and social life. We have made clear our concern with the "disconnected discussions" that reflect the ways that this discourse on the roles of ICTs in society takes place in the professional and academic literature. The problem that we see is that this discourse is almost independent of the rigorously and carefully accumulated body of knowledge that has arisen from the types of empirical research discussed throughout this book; even relatively sophisticated computer scientists can get swept up in entertaining but distracting debates about such topics as the likelihood that computers will develop superhuman intelligence.

We note, again, that beyond the simple-but-naïve veins of popular discourse on computing, there is reliable, evidence-based knowledge about the roles and effects of ICTs both in organizations and, more broadly, in social life. This body of knowledge, which we call Social Informatics, comes from more than thirty years of systematic, empirically anchored investigation, extensive analysis, and careful theorizing. As we have noted, this collected knowledge provides a rigorous but also rich and vivid basis for understanding the multiple roles that ICTs play in our lives. In doing this, we provide guidance for including Social Informatics findings and methods into the design and development of, policy-making processes for, educational curricula on, and scholarship in ICT.

References

AAPT. (1996). *Physics at the crossroads*. American Association of Physics Teachers. Retrieved from http://www.appt.org/programs/rupc/cross.html

Abraham, N., & Hoagland, K. E. (1995). *Shaping the future: New expectations for undergraduate education in science, mathematics, engineering and technology: Elaborating on the recommendations to and for SME&T faculty*. Washington, DC: Council on Undergraduate Research. Retrieved from http://www.cur.org/shaping.html

Abrams, M. (2002). A manual for policy analysts. GoJ-Cida capacity building project. Government of Jamaica.

Abran, A., & Moore, J. (2004). *Guide to the software engineering body of knowledge (SWEBOK)*. Los Alamitos, CA: IEEE Press.

Ackerman, M. (2000). The intellectual challenge of CSCW: The gap between social requirements and technical feasibility. *Human–Computer Interaction, 15*(2–3), 181–205.

Addison, T., & Vallabh, S. (2002). Controlling software project risks—An empirical study of methods used by experienced project managers. *Proceedings of SAICSIT 2002* (South African Institute of Computer Scientists and Information Technologists), 128–140.

Agre, P. (1996). Toward a critical technical practice: Lessons learned in trying to reform AI. In G. Bowker, L. Gasser, S. L. Star, and B. Turner (Eds.), *Bridging the great divide: Social science, technical networks, and cooperative work*. Mahwah, NJ: Lawrence Erlbaum.

Agre, P. (2000). Infrastructure and institutional change in the networked university. *Information, Communication, and Society* 3(4), 494–507.

Agre, P. (2003a). P2P and the promise of Internet equality. *Communications of the ACM, 46*(2), 39–42.

Agre, P. (2003b). Information and institutional change: The case of digital libraries. In A. P. Bishop, N. A. Van House, and B. P. Buttenfield (Eds.), *Digital library use: Social practice in design and evaluation*. Cambridge, MA: MIT Press.

Agre, P. E., & Schuler D. (Eds.). (1997). *Reinventing technology, rediscovering community: Critical studies in computing as a social practice*. Norwood, NJ: Ablex.

Agricultural Educator's Professional Development Program. (2003). Retrieved from http://www.ageds.iastate.edu/outreach/aepd/index.htm

Agricultural Extension Service of Nashville and Davidson County. (2003). Mission. Retrieved from http://www.nashville.org/aes/index.html

Allen, J. (2004). Redefining the network: Enrollment strategies in the PDA industry. *Information Technology & People, 17*(2), 171–185.

Anders, G., & Thurm, S. (1999). The innovators: The rocket under the tech boom: Heavy spending by basic industries. *Wall Street Journal* (Eastern edition). March 30.

Ashenhurst, R. L. (1972). Curriculum recommendations for graduate professional programs in information systems. *Communications of the ACM, 15*(5), 363–398.

Attewell, P. (1996). Information technology and the productivity challenge. In R. Kling (Ed.), *Computerization and controversy*. San Diego: Academic Press.

Attewell, P. (1998). *Research on information technology impacts*. Retrieved from http://www.nap.edu/readingroom/books/esi/appb.html#attewell

Avgerou, C. (2002). *Information systems and global diversity*. Oxford: Oxford University Press.

Avison, D., & Wood-Harper, T. (2003). *Bringing social and organisational issues into information systems development: The story of multiview, socio-technical and human cognition elements of information systems*. Hershey, PA: Idea Group Publishing.

Bangemann, M. et al. (1995). *Europe and the global information society: Recommendations to the European Council*. Retrieved from http://www.medicif.org/Dig-library/ECdocs/reports/Bangemann.htm

Barley, S. R. (1986). Technology as an occasion for structuring: Evidence from observation of CT scanners and the social order of radiology departments. *Administrative Science Quarterly, 31*, 78–108.

Barley, S. R. (1988). On technology, time and the social order: Technically induced change in the temporal order of radiological work. In F. Dubinskas (Ed.), *Making time: Ethnographies of high-tech organizations*. Philadelphia: Temple University Press.

Barley, S. R. (1990). The alignment of technology and structure through roles and networks. *Administrative Science Quarterly, 35*, 61–103.

Barley, S. R. (1996). Technicians in the workplace: Ethnographic evidence for bringing work into organization studies. *Administrative Science Quarterly, 41*, 404–441.

Barton, D., & Hamilton, M. (2000). Literacy practices. In D. Barton, R. Ivanic, & Mary Hamilton (Eds.), *Situated literacies: Reading and writing in context* (pp. 6–17). New York: Routledge.

Bauer, J. (2001). Technology policy and democratic societies. *Journal of Communication, 51*(2), 413–418.

Beath, C., & Orlikowski, W. (1994). The contradictory structure of systems development methodologies: Deconstructing IS-user relationships in information engineering. *Information Systems Research, 5*(4), 350–377.

Belanger, Y. (2000). *Laptop computers in the K-12 classroom.* ERIC Digest. ED440644. Syracuse, NY: ERIC Clearinghouse on Information and Technology.

Bell, D. (1973). *The coming of post-industrial society: A venture in social forecasting.* New York: Basic Books.

Bell, D. (1980). The social framework of the information society. In T. Forester (Ed.), *The microelectronics revolution: The complete guide to the new technology and its impact on society* (pp. 500–549). Cambridge, MA: MIT Press.

Benbasat, I., & Zmud, R. (1999). Empirical research in information systems: The practice of relevance. *MIS Quarterly, 23*(1), 3–16.

Benton Foundation. (1996). *A nation of opportunity: A final report of the United States Advisory Council on the National Information Infrastructure.* Retrieved from http://www.benton.org/Library/KickStart/nation_summary.html

Bimber, B. (1996a). *The death of an agency: Office of Technology Assessment & budget politics in the 104th Congress.* Retrieved from http://www.english.ucsb.edu/faculty/ayliu/research/bimber.html

Bimber, B. (1996b). *The politics of expertise in Congress: The rise and fall of the Office of Technology Assessment.* New York: State University of New York Press.

Bishop, A., & Star, S. L. (1996). Social informatics for digital libraries. *Annual Review of Information Science and Technology (ARIST), 31,* 301–403.

Bowker, G., & Star, S.L. (1999) *Sorting things out: Classification and its consequences.* Cambridge, MA: MIT Press.

Braa, J., & Monteiro, E. (1996). Infrastructure and institutions: The case of public health in Mongolia. In E. M. Roche and M. J. Blaine (Eds.), *Information technology, development and policy. Theoretical perspectives and practical challenges* (pp. 171–188). London: Avebury.

Brooks, F. P. J. (1987). No silver bullet: Essence and accidents of software engineering. *Computer, 20*(4), 10–19.

Brooks, F. P. J. (1996). The computer scientist as toolsmith II. *Communications of the ACM, 39*(3), 61–68.

Brown, J. S., & Duguid P. (1995). The social life of documents. *Release 1.0: Esther Dyson's Monthly Report, 10–95,* 1–23.

Brown, J. S., & Duguid, P. (2000). *The social life of information.* Boston: Harvard Business School Press.

Bureau of Labor Statistics. (1998). *Employment projections: Fastest growing occupations, 1996–2006.* Retrieved from http://stats.bls.gov/emptab1.htm

Bureau of Labor Statistics. (2002). *Labor force statistics from the current population survey.* U.S. Department of Labor. Retrieved from http://data.bls.gov/cgi-bin/surveymost

Bureau of Labor Statistics. (2004a). *BLS releases 2002–12 employment projections.* Retrieved from ftp://ftp.bls.gov/pub/news.release/History/ecopro.02112004.news

Bureau of Labor Statistics. (2004b). *Occupational outlook handbook: Computer and information systems managers.* Retrieved from http://www.bls.gov/oco/ocos258.htm

Burkhardt, M. (1994). Social interaction effects following a technological change: A longitudinal investigation. *Academy of Management Journal, 37*(4), 869–898.

Bush, V. (1945). As we may think. *The Atlantic Monthly, 176*(1), 101–108. Retrieved from http://www.theatlantic.com/unbound/flashbks/computer/bushf.htm

Butler, T. (2003). An institutional perspective on developing and implementing intranet- and Internet-based information systems. *Information Systems Journal, 13,* 209–231.

Cahoon, J. (2003). Must There Be So Few? Including Women in CS? *25th International Conference on Software Engineering, May 3-10, Portland, Oregon.* IEEE Press, 668-675.

Carmel, E. (1997). American hegemony in packaged software trade and the "culture" of software. *The Information Society, 13,* 125–142.

Carmel, E., & Agarwal, R. (2002). The maturation of offshore sourcing of information technology work. *MISQ Executive, 1*(2), 65–77.

Carr, A. (1997). User-design in the creation of human learning systems. *Educational Technology Research & Development, 45*(3), 5–23.

Carr, N. (2004). Burned by IT. *Industrial Engineer, 36*(8), 28–33.

Chartier, R. (1994). *The order of books: Readers, authors, and libraries in Europe between the fourteenth and eighteenth centuries.* Trans. Lydia G. Cochrane. Stanford, CA: Stanford University Press.

Checkland P., & Scholes J. (1990). *Soft systems methodology in action.* Chichester, UK: John Wiley & Sons.

Choo, C. W. (2001). Environmental scanning as information seeking and organizational learning. *Information Research, 7*(1). Retrieved from http://informationr.net/ir/7-1/paper112.html

Ciborra, C. (Ed.). (1996). *Groupware and teamwork: Invisible aid or technical hindrance?* New York: John Wiley.

Christie, J. (1998). *Texas board of education: Paving the way for electronic textbooks. Institute for Cyber Information.* Retrieved from http://www.futureprint.kent.edu/articles/christie01.htm

Clement, A. (1994). Computing at work: Empowering action by "low-level users." *Communications of the ACM, (37)*1, 52–65.

Clement, A., & Halonen, C. (1998). Collaboration and conflict in the development of a computerized dispatch facility. *Journal of the American Society for Information Science, 49,* 1090–1100.

Clinton, W. J., & Gore, A. (1997). A framework for global electronic commerce. Washington, DC: USGRP. Retrieved from http://www.technology.gov/digeconomy/framewrk.htm

Coley, R. J., Cradler, J., & Engel, P. K. (1997). *Computers and classrooms: The status of technology in U.S. schools.* Princeton, NJ: Educational Testing Service. Retrieved from http://198.138.177.34/research/pic/compclass.html

Commission on Behavioral and Social Sciences and Education (CBASSE). (1994). *Organizational linkages: Understanding the productivity paradox.* Washington, DC: National Academy Press. Retrieved from http://www.nap.edu/catalog/2135.html

Computer Science Accreditation Committee (CSAC). (1998). *Guidance for accrediting programs in computer science in the United States.* Version 0.4, February 14, Computing Sciences Accreditation Board. Retrieved from http://www.csab.org/guidance2k_v04.html

Computer Science Accreditation Committee (CSAC). (2000). *Criteria for accrediting programs in computer science in the United States,* (January, Version 1.0). Retrieved from http://www.csab.org/criteria2k_v10.html

Computer Science and Telecommunications Board (CSTB). (2004). About CSTB: CSTB's impact. Retrieved from http://www7.nationalacademies.org/cstb/impact.html

Computer Science and Telecommunications Board (CSTB), Commission on Physical Sciences, Mathematics, and Applications (CPSMA), National Research Council. (1994). *Information technology in the service society: A twenty-first century lever.* Washington, DC: National Academy Press. Retrieved from http://www.nap.edu/catalog/2237.html

Computer Science and Telecommunications Board (CSTB), Commission on Physical Sciences, Mathematics, and Applications (CPSMA), National Research Council. (1997a). *More than screen deep: Toward every-citizen interfaces to the nation's information infrastructure.* Washington, DC: National Academy Press. Retrieved from http://www.nap.edu/readingroom/books/screen

Computer Science and Telecommunications Board (CSTB), Commission on Physical Sciences, Mathematics, and Applications (CPSMA), National Research Council. (1997b). *Defining a decade: Envisioning CSTB's second 10 years.* Washington, DC: National Academy Press. Retrieved from http://www.nap.edu/catalog/5903.html

Computer Science and Telecommunications Board (CSTB), National Research Council. (1998). Design and Evaluation: A review of the state-of-the-art. *D-Lib Magazine,* (July/August). Retrieved from http://www.dlib.org/dlib/july98/nrc/07nrc.html

Cooper, M. N. (2002). *Does the digital divide still exist? Bush Administration shrugs, but evidence says "Yes."* Washington, DC: Consumer Federation of America.

Copeland, D., Mason, R., & McKenney, J. (1995). SABRE: The development of information-based competence and execution of information-based competition. *IEEE Annals of the History of Computing, 17*(3), 30 -57.

Crabtree, A., Nichols, D. M., O'Brien, J., Rouncefield, M., & Twidale, M. B. (2000). Ethnomethodologically-informed ethnography and information system design. *Journal of the American Society of Information Science and Technology, 51*(7), 666–682.

Cranor, L. F., & Greenstein, S. (2003). Introduction. In L. F. Cranor & S. Greenstein (Eds.), *Communications policy and information technology.* Cambridge, MA: MIT Press. Reprint available from http://www.acm.org/ubiquity/book/l_cranor_1.html

Cronin, B. (1999). Will e-publishing save academic publishing? *Chronicle of Higher Education,* October 15, A25.

Dalcher, D., & Drevin, L. (2003). Learning from information systems failures by using narrative and antenarrative methods. In J. Eloff, P. Kotzé, A. Engelbrecht, & M. Eloff (Eds.), *Proceedings of SAICSIT 2003* (South

African Institute of Computer Scientists and Information Technologists), 137–142.

Danziger, J. N., Dutton, W. H., Kling, R., & Kraemer, K. L. (1982). *Computers and politics: High technology in American local governments.* New York: Columbia University Press.

Davenport, T. H. (1996). Software as socialware. *CIO Magazine On Line.* Retrieved from http://www.cio.com/archive/030196_dave_content. html

Davenport, T. H. (1997). Storming the ivory tower. *CIO Magazine,* April 15. Retrieved from http://www.cio.com/archive/041597_think_content. html

Davenport, T. H. (1998). Putting the enterprise into the enterprise system. *Harvard Business Review, 76*(4), 121–134.

Davenport, T. H. (2000). *Mission critical.* Cambridge, MA: Harvard Business School Press.

Davis, G. B., Gorgone, J. T., Couger, J. D., Feinstein, D. L., & Longenecker, H. E. J. (1997). *IS '97: Model curriculum and guidelines for undergraduate degree programs in information systems.* Association for Computing Machinery (ACM), Association for Information Systems (AIS), Association for Information Technology Professionals (AITP).

Denning, P. J., & Dargan, P. A. (1994). A discipline of software architecture. *ACM interactions, 1*(1), 55-65.

Dertouzous, M., & Gates, B. (1998). *What will be: How the new world of information will change our lives.* San Francisco: Harper.

Diffie, W., & Landau, S. (1998). *Privacy on the line: The politics of wiretapping and encryption.* Cambridge, MA: The MIT Press.

Donaldson, J. (2004). Analyzing systems failures through the use of case histories. *Proceedings of the 2004 International Symposium on Empirical Software Engineering* (ISESE '04). Los Alamitos, CA: IEEE Computer Society.

Donnermeyer, J. F., & Hollifield, C. A. (2003). Digital divide evidence in four rural towns. *IT & Society, 1*(4), 107–11.

Downs, A. (1967). A realistic look at the payoffs from urban data systems. *Public Administration Review, 27*(3), 204–210.

Drummond, H., & Hodgson, J. (2003). The chimpanzees' tea party: A new metaphor for project managers. *Journal of Information Technology, 18*(3), 151–159.

Dutton, W. H. (Ed.). (1996). *Information & communication technologies: Vision & realities.* New York: Oxford University Press.

Dutton, W. H. (1999). *Society on the line: Information politics in the digital age.* Oxford and New York: Oxford University Press.

Dutton, W. H., & Kraemer, K. L. (1985). *Modeling as negotiating: The political dynamics of computer models in the policy process.* Norwood, NJ: Ablex.

Erikson, T. (1997). *Supporting interdisciplinary design: Towards pattern languages for workplaces.* Retrieved from http://www.pliant.org/personal/Tom_Erickson/Patterns.Chapter.html

Eschenfelder, K. (2004). The customer is always right, but whose customer is more important? Conflict and Web site classification schemes. *Information Technology & People, 16*(4), 419–439.

Faraj, S., Kwon, D., & Watts, S. (2004). Contested artifact: Technology sensemaking, actor networks, and the shaping of the Web browser. *Information Technology & People, 17*(2), 186–209.

Feldman, M. S. (1989). *Order without design: Information production and policy making.* Stanford, CA: Stanford University Press.

Fitzduff, M. (2000). *From shelf to field: Functional knowledge for conflict management.* Presented at Facing Ethnic Conflicts: Perspectives from research and policy making, Bonn, December 14–16. Retrieved from http://www.incore.ulst.ac.uk/home/policy/policy/Shelf_to_Field_3.pdf

Fleck, J. (1994). Learning by trying: The implementation of configurational technology. *Research Policy, 23,* 637–652.

Fletcher, P. D., Rollier, B., Small, R. V., & Wildemuth, B. M. (1995). Lifelong learning for information systems professionals. Retrieved from http://hsb.baylor.edu/ramsower/acis/papers/fletcher.htm

Forsythe, D. (1992). Blaming the user in medical informatics. *Knowledge and society: The anthropology of science and technology, 9,* 95–111.

Forsythe, D. (1994). Engineering knowledge: The construction of knowledge in artificial intelligence. *Social Studies of Science, 24,* 105–113.

Fountain, J. (2001). *Building the virtual state: Information technology and institutional change.* Washington, DC: Brookings Institute.

Frauenheim, E. (2004). Study: IT spending looking better. C|Net News.com. Retrieved March 29 from http://att.com/Study:+IT+spending+looking+better/2100-1012_3-5181480.html

Freeman, P., & Aspray, W. (1999). *The supply of information technology workers in the United States.* Washington, DC: Computing Research Association. Retrieved from http://www.cra.org/reports/wits

Friedman, W. H. (2002). *The digital divide.* Seventh Americas Conference on Information Systems, 2081–2087.

Friedman, B., & Nissenbaum, H. (1996). Bias in computer systems. *ACM Transactions on Information Systems, 14*(3), 330–347.

Fulk, J., & DeSanctis, G. (1998). Articulation of communication technology and organizational form. In G. DeSanctis & J. Fulk (Eds.), *Shaping organizational form: Communication, connection, and community.* Newbury Park, CA: Sage.

Gal-Ezer, J., & Harel, D. (1998). What (else) should CS educators know? *Communications of the ACM, 41*(9), 77–84.

Garfinkel, S. (2000). *Database nation.* San Fransisco: O'Reilly & Associates.

Gasser, L. (1986). An empirical study of the integration of computing into routine work. *Proceedings of the 1986 ACM Conference on Office Information Systems*, Providence, RI, October, 1986.

Gasson, S. (1998). Framing design: A social process view of information system development. *Proceedings of ICIS '98*, Helsinki, Finland, December.

Gates, W. (1995). *The road ahead.* New York: Viking.

General Accounting Office (GAO). (2004). *Workforce challenges and opportunities for the 21st century: Changing labor force dynamics and the role of government policies.* U.S. Washington, DC: GAO Report GAO-05-845SP.

George, J. (2003). *Social issues of computing.* New York: Oxford.

Gibbs, W. W. (1997). Taking computers to task: Coming generations of computers will be more fun and engaging to use. But will they earn their keep in the workplace? *Scientific American*, July. Retrieved from http://www.sciam.com/0797issue/0797trends.html

Goodman, P. S., Sproull, L. S., & Associates. (1990). *Technology and Organizations.* San Francisco: Jossey-Bass.

Gosling, J., & Mintzberg, H. (2004). The education of practicing managers. *MIT Sloan Management Review, 45*(4), 19–23.

Grant, S. G. (2000). Teachers and tests: Exploring teachers' perceptions of changes in the New York State testing program. *Education Policy Analysis Archives, 8*(14), February 24. Retrieved from http://olam.ed.asu.edu/epaa/v8n14.html

Greenbaum, J., & Kyng, M. (1991). Situated design. In J. Greenbaum & M. Kyng (Eds.), *Design at work: Cooperative design of computer systems* (pp. 1–24). Hillsdale, NJ: Lawrence Erlbaum.

Grover, M. B. (1999). E-commerce, e-pain: So you wanna sell on-line? Get ready for headaches. *Forbes Magazine, 163*(5), March 8, 124–128.

Grudin, J. (1991). CSCW: Introduction to the special section. *Communications of the ACM, 34*(12), 30-34.

Guevara, K., & Ord, J. (1996). The search for meaning in a changing work context. *Futures, 28*(8), October, 709–722.

Guinan, P., Cooprider, J., & Sawyer, S. (1997). The effective use of automated application development tools: A four-year longitudinal study of CASE. *IBM Systems Journal, 38*, 124–141.

Guzdial, M., & Weingarten, F. W. (1995). *Setting a computer science research agenda for educational technology.* National Science Foundation & Georgia Institute of Technology.

Hara, N., & Kling, R. (2002). Students' difficulties in a Web-based distance education course: An ethnographic study. In W. H. Dutton & B. D. Loader (Eds.), *Digital academe: New media and institutions in higher education and learning* (pp. 62–84). London: Taylor & Francis/Routledge.

Harris, D. H. (Ed.). (1994). *Organizational linkages: Understanding the productivity paradox.* Washington, DC: National Academy Press.

Hartmanis, J., & Lin, H. (Eds.). (1992). *Computing the future: A broader agenda for computer science and engineering.* Washington, DC: National Academy Press.

Hayes, N. (1999). *Safe enclaves, political enclaves and knowledge working.* Critical Management Studies Conference, July 14–16, Manchester, England. Retrieved from http://www.mngt.waikato.ac.nz/ejrot/cmsconference/papers_infotech.htm

Haythornthwaite, C., & Hagar, C. (2004). The social worlds of the Web. In B. Cronin (Ed.), *Annual Review of Information Science and Technology, 39* (pp. 311–346). Medford, NJ: Information Today.

Heeks, R. (2002). Information systems and developing countries: Failure, success, and local improvisations. *The Information Society, 18*, 101–112.

High Level Expert Group on the Social and Societal Aspects of the Information Society (HLEG). (1997). *Building the european information society for us all: Final policy report of the High Level Expert Group.* European Commission, DG V/B-4. Retrieved from http://www2.bologna.enea.it/MIDAS-NET/library/buildeis.html

Hirschheim, R. A. (1986). The effect of a priori views on the social implications of computing: The case of office automation. *Computing Surveys, 18*(2), 165–195.

Hodas, S. (1996). Technology refusal and the organizational culture of schools. In R. Kling (Ed.), *Computerization and controversy* (pp. 197–218). San Diego: Academic Press.

Hoffman, J. (1999). Writers, texts and writing acts: Gendered user images in word processing software. In D. MacKenzie & D. Wajcman (Eds.),

Secondary writers, texts and writing acts: Gendered user images in word processing software (pp. 222–243). Buckingham, UK: Open University Publishers.

Horton, K., Davenport, E., & Wood-Harper, A. (2005). Exploring sociotechnical interaction with Rob Kling: Five big ideas. *Information Technology & People, 18*(1), 50–67.

Howard, P., Rainie, L., & Jones, S. (2002). Days and nights on the Internet. In B. Wellman & C. Haythornthwaite (Eds.), *The Internet in everyday life* (pp. 45–73). Oxford, UK: Blackwell.

Howcroft, D., & Wilson, M. (1999). Paradoxes of participatory design: The end-user perspective. *Critical Management Studies Conference, July 14–16, Manchester, England.* Retrieved from http://www.mngt.waikato. ac.nz/ejrot/cmsconference/papers_infotech.htm

Huff, C., & Finholt, T. (1994). *Social issues in computing: Putting computing in its place.* New York: McGraw-Hill.

Huff, C. W., & Martin, D. (1995). Computing consequences: A framework for teaching ethical computing. *Communications of the ACM, 38*(12), 75–84.

IBM. (2004). NGI: Next Generation Internet. Retrieved from http://www.ngi. ibm.com

Iowa State University. (1996). *Agricultural extension education: A career that works.* Retrieved from http://www.ag.iastate.edu/departments/aged/ undergra/extensio.htm

Information Society Promotion Office (ISPO), European Commission. (2000). e*Europe—An information society for all.* Retrieved from http://www.ispo.cec.be/policy/i_europe.html

Information Technology Association of America (ITAA). (1997). *Help wanted: The IT workforce gap at the dawn of a new century.* Arlington, VA: ITAA.

Information Technology Association of America (ITAA). (2001). *When can you start? Building better information technology skills and careers.* Arlington, VA: ITAA.

Internet2. (2004). *Applications.* Retrieved from http://apps.internet2.edu

The Internet Systems Consortium. (2003). *Internet domain survey*, January, 2003. Retrieved from http://www.isc.org/index.pl?/ops/ds

Introna, L., & Nissenbaum, H. (2000). Shaping the Web: Why the politics of search engines matter. *Information Society, 16*(3), 1–17.

James, G. (1997). IT fiascoes … and how to avoid them. *Datamation,* November. Retrieved from http://datamation.earthweb.com/ cio/11disas.html

Jamison, A., & Rohracher, H. (2001). Introduction: Technology studies and sustainable development. *Technology Analysis & Strategic Management, 13*(1), 5–8.

Johnson, B., & Rice, R. E. (1987). *Managing organizational innovation: The evolution from word processing to office information systems.* New York: Columbia University Press.

Joy, W. (2000). Why the future doesn't need us. *Wired, 8*(4), April, 238–262.

Kanter, R. M. (2004). *Confidence: How winning streaks and losing streaks begin and end.* New York: Crown Business.

Karoly, L., & Panis, C. (2004). *The 21st century at work: Forces shaping the future workforce and workplace in the United States.* Pittsburgh: Rand Corporation.

Katzer, J., & Fletcher, P. T. (1992). The information environment of managers. In M. E. Williams (Ed.), *Annual Review of Information Science and Technology* (pp. 227–263). White Plains, NY: Knowledge Industry Publications.

Keefe, D., & Zucker, A. (2003). Ubiquitous computing projects: A brief history. Ubiquitous Computing Evaluation Consortium, April. SRI Project P12269. NSF Grant No. REC-0231147. Retrieved from http://www.ubiqcomputing.org/Overview.pdf

Keil, M., & Carmel, E. (1995). Customer–developer links in software development. *Communications of the ACM, 37*(5), 33–44.

Khattri, N., Reeve, A. L., & Adamson, R. J. (1997). *Assessment of student performance: Studies of education reform.* Office of Educational Research and Improvement. Washington, DC: U.S. Department of Education. Retrieved from http://www.ed.gov/pubs/SER/ASP/stude.html

Kiesler, S. (Ed.). (1997). *Culture of the internet.* Mahwah, NJ: Lawrence Erlbaum.

King, J. L. (1983). Centralized versus decentralized computing: Organizational considerations and management options. *Computing Surveys, 15*(4), 320–349.

King, J. L. (1997). Cross-disciplinary, social-context research. In Computer Science and Telecommunications Board (CSTB), Commission on Physical Sciences, Mathematics, and Applications (CPSMA), National Research Council, *More than screen deep: Toward every-citizen interfaces to the nation's information infrastructure* (pp. 411–416). Washington, DC: National Academy Press. Retrieved from http://www.nap.edu/readingroom/books/screen

Kirkpatrick, D. (1993). Groupware goes boom. *Fortune, 27*(12), 93.

Kling, R. (1977). The organizational context of user-centered software design. *MIS Quarterly, 1*(4), 41–52.

Kling, R. (1980). Social issues and impacts of computing: From arena to discipline. *Proceedings from the Second Conference on Computers and Human Choice,* Vienna, June 1979. Amsterdam: North Holland Publishing Company.

Kling, R. (1983). Value conflicts in the deployment of computing applications: Cases in developed and developing countries. *Telecommunications Policy, 7*(1), 12–34.

Kling, R. (1991). Computerization and social transformations. *Science, Technology, and Human Values, 16*(3), 342–267.

Kling, R. (1992). Behind the terminal: The critical role of computing infrastructure in effective information systems' development and use. In W. Cotterman & J. Senn (Eds.), *Challenges and strategies for research in systems development* (pp. 153–201). New York: John Wiley.

Kling, R. (1993). Organizational analysis in computer science. *The Information Society, 9*(2), 71–87.

Kling, R. (1994). Organizational analysis in computer science. *The Information Society, 9*(2), (Mar–Jun, 1993), 71–87. Revised April, 1994 (v. 13.3). Retrieved from http://www.slis.indiana.edu/~kling/home.html

Kling, R. (1996a). *Content and pedagogy in teaching about the social aspects of computerization.* Bloomington, IN: Center for Social Informatics, School of Library and Information Science, Indiana University.

Kling, R. (1996b). Information and computer scientists as moral philosophers and social analysts. In R. Kling (Ed.), *Computerization and controversy: Value conflicts and social choices, 2nd ed.* (pp. 32–38). San Diego: Academic Press.

Kling, R. (1996c). Systems safety, normal accidents, and social vulnerability. In R. Kling (Ed.), *Computerization and controversy: Value conflicts and social choices, 2nd ed.* (pp. 315–327). San Diego: Academic Press.

Kling, R. (1996d). *Computerization and controversy: Value conflicts and social choices, 2nd ed.* San Diego: Academic Press.

Kling, R. (1999a). What is social informatics and why does it matter? *D-Lib Magazine, 5*(1). Retrieved from http://www.dlib.org:80/dlib/january99/kling/01kling.html

Kling, R. (1999b). Can the 'Next Generation Internet' effectively support ordinary citizens? *The Information Society, 15*(1), 57–63.

Kling, R. (2000). Learning about information technologies and social change: The contribution of Social Informatics. *The Information Society, 16*(3), 212–234.

Kling, R. (2002). Critical professional discourses about information and communications technologies and social life in the U.S. In K. Brunnstein & J. Berleur (Eds.), *Human choice and computers: Issues of choice and quality of life in the information society* (pp. 1–20). New York: Kluwer Academic.

Kling, R. (2003). Critical professional education about information and communications technologies and social life. *Information Technology & People, 16*(4), 394–418.

Kling, R., & Iacono, C. S. (1984). Computing as an occasion for social control. *Journal of Social Issues, 40*(3), 77–96.

Kling, R., & Iacono, C. S. (1988). The mobilization of support for computerization: The role of computerization movements. *Social Problems, 35*(3), 226–243.

Kling, R., & Iacono, C. S. (1995). Computerization movements and the mobilization of support for computerization. In S. L. Star (Ed.), *Ecologies of knowledge: Work and politics in science and technology* (pp. 119–153). New York: SUNY Press.

Kling, R., & Jewett, T. (1994). The social design of worklife with computers and networks: An open natural systems perspective. In M. Yovits (Ed.), *Advances in computers*, vol. 39 (pp. 239–293). Orlando, FL: Academic Press.

Kling, R., & Lamb, R. (2000). IT and organizational change in digital economies: A socio-technical approach. In B. Kahin & E. Brynjolfsson (Eds.), *Understanding the digital economy—Data, tools and research.* Cambridge, MA : MIT Press.

Kling, R., McKim, G., & King. (2001). A bit more to IT: Scholarly communication forums as socio-technical interaction networks. Retrieved from http://www.slis.indiana.edu/CSI/WP/wp01-02B.html

Kling, R., Rosenbaum, H., & Hert, C. A. (Eds.). (1998). *Special Issue on Social Informatics. Journal of the American Society for Information Science, 49*(12). Retrieved from http://www.asis.org/Publications/JASIS/v49n1298.html

Kling, R., & Scacchi, W. (1982). The Web of computing: Computer technology as social organization. *Advances in Computers, 21*, 1–90.

Kling, R., & Star, L. (1998). Human centered systems in the perspective of organizational and social informatics. *Computers and Society, 28*(1). Retrieved from http://www.slis.indiana.edu/kling/pubs/kling9801.pdf

Kling, R., Star, S. L., Kiesler, S., Agre, P., Bowker, G., Attewell, P., & Ntuen, C. (1997). *Human centered systems in the perspective of social informatics.* Bloomington, IN: Indiana University.

Kling, R., & Tillquist, J. (1996). *Computerization, work values, and changing worklife.* Final Report for NSF Grant SRB-93-21375.

Kling, R., & Zmuidzinas, M. (1994). Technology, ideology and social transformation: The case of computing and work organization. *Revue International de Sociologie, 2*(3), 28–56.

Kraut, R., Dumais, S., & Koch, S. (1989). Computerization, productivity, and the quality of work-life. *Communications of the ACM, 32*(2), 220–228.

Kubicek, H., Dutton, W. H., and Williams, R. (Eds.). (1997). *The social shaping of information superhighways: European and American roads to the information superhighway.* New York: Campus Verlag/St. Martin's Press.

Kurzweil, R. (1999). *The age of spiritual machines: When computers exceed human intelligence.* New York: Penguin.

Lamb, R. (1997). *Interorganizational relationships and information services: How technical and institutional environments influence data gathering practices.* Unpublished doctoral thesis, University of California, Irvine.

Lamb, R., & Kling, R. (2003). Reconceptualizing users as social actors in information systems research. *MIS Quarterly, 27*(2), 197–235.

Lamb, S., & Sawyer, S. (2005). On extending Social Informatics from a rich legacy of networks and conceptual resources. *Information Technology & People, 18*(1), 9–20.

Landauer, T. (1995). *The trouble with computers.* Cambridge, MA: MIT Press.

Lang, M. (2003). Communicating academic research findings to IS professionals: An analysis of problems. Special series: Informing each other. *Informing Science, 6*, 9p. Retrieved from http://inform.nu/Articles/Vol6/v6p021-029.pdf

Large, A. (2004). Children, teenagers and the Web. In B. Cronin (Ed.), *Annual Review of Information Science and Technology, 39* (pp. 347–392). Medford, NJ: Information Today.

Large Scale Networking Working Group of the Computing, Information, and Communications R&D Subcommittee. Next Generation Internet [NGI] Initiative. (1996). NGI Concept Paper. NGI Publications.

Laudon, K. C. (1974). *Computers and bureaucratic reform: The political functions of urban information systems.* New York: John Wiley.

Laudon, K. C. (1986). *Dossier society: Value choices in the design of national information systems.* New York: Columbia University Press.

Laudon, K. C., & Marr, K. L. (1995). *Information Technology and Occupational Structure.* AIS Americas Conference on Information Systems. Retrieved from http://hsb.baylor.edu/ramsower/acis/papers/laudon.htm

Leavitt, H. J., & Whisler, T. L. (1958). Management in the 1980s. *Harvard Business Review,* November–December, 41–48.

Lee, H., & Sawyer, S. (2002). Conceptualizing time and space: Information technology, work and organization. In B. Galliers, L. Applegate, & J. DeGross (Eds.), *Proceedings of the 2002 International Conference on Information Systems,* December 15–18. Barcelona, Spain: ACM Press.

Leonard-Barton, D. (1988). Implementation as a mutual adaptation of technology and organization. *Research Policy, 17,* 251–267.

Levine, H. G., & Rossmoore, D. (1993). Diagnosing the human threats to information technology implementation: A missing factor in systems analysis illustrated in a case study. *Journal of Management Information Systems, 10*(2), 55–73.

Lidtke, D. K. (1997). *President's message from 1997 annual report.* Stanford, CT: Computing Sciences Accreditation Board.

Liker, J. K., Roitman, D. B., & Roskies, E. (1987). Changing everything all at once: Work life and technological change. *Sloan Management Review,* Summer, 29–47.

Lucas, Jr., H. C. (1975). *Why information systems fail.* New York: Columbia University Press.

Mabry, L. (1999). Writing to the rubric: Lingering effects of traditional standardized testing on direct writing assessment. *Phi Delta Kappan, 80*(9), 673–679. Retrieved from http://www.pdkintl.org/kappan/kmab9905.htm

MacKenzie, D., & Wajcman, J. (1999). *The social shaping of technology.* (2nd ed.). Philadelphia: Open University Press.

Mann, J. (2002). IT education's failure to deliver successful information systems: Now is the time to address the IT-user gap. *Journal of Information Technology Education, 1*(4), 15. Retrieved from http://jite.org/indexes/Vol1Issue4.htm

Margherio, L., Henry, D., Cooke, C., & Montes, S. (1998). *The emerging digital economy.* Washington, DC: U.S. Department of Commerce. Retrieved from http://www.ecommerce.gov/emerging.htm

Markus, M. L. (1981). Implementation politics: Top management support and user involvement. *Systems, Objectives, Solutions, 1*(4), 203–215.

Markus, M. L. (1983). Power, politics, and MIS implementation. *Communications of the ACM, 26*(6), 430–444.

Markus, M. L. (1994). Finding a 'happy medium': Explaining the negative effects of electronic communication on social life at work. *ACM Transactions on Information Systems, 12*(2), 119–149.

Markus, M., Axline, S., Petrie, D., & Tanis, C. (2000). Learning from adopters' experiences with ERP: Problems encountered and success achieved. *Journal of Information Technology, 15*(4), 245–266.

Markus, M., Beath, C., & Silver, M. (1995). The information technology inter-action model: A model for the MBA core course. *MIS Quarterly, 19*(3), 361–390.

Markus, M., & Bjorn-Andersen, N. (1987). Power over users: Its exercise by system professionals. *Communications of the ACM, 30*(6), 498–504.

Markus, M. L., & Benjamin, R. I. (1997). The magic bullet theory in IT-enabled transformation. *Sloan Management Review, 38*(2), 55–68.

Markus, M. L., & Keil, M. (1994). If we build it, they will come: Designing information systems that users want to use. *Sloan Management Review, 35*(4), 11–25.

Martin, C. D., Huff, C., Gotterbarn, D., & Miller, K. (1996). Implementing a tenth strand in the CS curriculum. *Communications of the ACM, 39*(12), 75–84.

McDermott, L. C. (1993). How we teach and how students learn—A mis-match? *American Journal of Physics, 61*(4), 295–298.

McDermott, L. C., & Redish, E. F. (1999). Resource letter: PER-1: Physics education research. *American Journal of Physics, 67*(10), 755–767.

McKeen, J. D., & Smith, H.A. (1997). *Management challenges in IS: Successful strategies and appropriate action.* Chichester, UK: John A. Wiley and Sons.

McKenna, P. G., & Barber, W. G., Jr. (1987). Extension goes to high school. *Journal of Extension, 25*(4). Retrieved from http://joe.org/joe/1987winter/a4.html

McKenzie, J. (2001). The unwired classroom: Wireless computers come of age. *From Now On: The Educational Technology Journal, 10*(4). Retrieved from http://www.fno.org/jan01/wireless.html

Meltsner, A. J. (1976). *Policy analysts in the bureaucracy.* Berkeley, CA: University of California Press.

Merton, R. (1936). The unanticipated consequences of purposive social action. *American Sociological Review, 1*, 894–904.

Michigan State University. (2002). *Agricultural Extension Education at Michigan State University*. Retrieved from http://www.canr.msu.edu/aee/background.html

Miyamoto, S., Midorikawa, N., & Nakayama, K. (1990). A view of studies on bibliometrics and related subjects in Japan. In C. L. Borgman (Ed.), *Scholarly communication and bibliometrics* (pp. 73–83). Newbury Park, CA: Sage.

Montealegre, R., & Keil, M. (2000). De-escalating information technology projects: Lessons from the Denver International Airport. *Management Information Sciences Quarterly, 24*(3), 417–447.

Moody, D. L. (2000). Building links between IS research and professional practice: Improving the relevance and impact of IS research. In *Proceedings of International Conference on Information Systems* (pp. 351–360) ICIS, 2000, December 10–13, Brisbane, Australia.

Moravec, H. P. (1999). *Robot: Mere machine to transcendent mind*. New York: Oxford University Press.

Mowshowitz, A. (2002). *Virtual organization: Toward a theory of societal transformation stimulated by information technology*. New York: Quorum Books.

Mumford, E. (2000). Socio-technical design: An unfulfilled promise or a future opportunity? In R. Baskerville, J. Stage, & J. DeGross (Eds.), *Organizational and Social Perspectives on Information Technology* (pp. 33–46). IFIP TC8 WG8.2 International Working Conference on the Social and Organizational Perspective on Research and Practice in Information Technology, June, Aalborg, Denmark.

Nadler, G. (1963). *Work Design*. Homewood, IL: Richard D. Irwin.

National Academy. (1996). *From analysis to action: Undergraduate education in science, mathematics, engineering, and technology*. Retrieved from http://www.nap.edu/readingroom/records/NI000012.html

National Coordination Office for Computing Infomation, and Communications. (1998). *Information technology: Transforming our society*. President's Information Technology Advisory Committee.

National Science Foundation (NSF). (1996). *Shaping the future: New expectations for undergraduate education in science, mathematics, engineering and technology*, National Science Foundation Division of Undergraduate Education, Arlington, VA. Retrieved from http://www.ehr.nsf.gov/EHR/DUE/documents/review/96139/start.htm

National Science Foundation (NSF). (2002). *Science and engineering indicators*. National Science Foundation, Arlington, VA. Retrieved from http://www.nsf.gov/sbe/srs/seind02/toc.htm

National Telecommunications and Information Administration (NTIA). (1998). *Falling through the Net II: New data on the digital divide*. Washington, DC: Author. Retrieved from http://www.ntia.doc.gov/ntiahome/net2/falling.html

National Telecommunications and Information Administration (NTIA). (1999). *Falling through the Net: Defining the digital divide*. Washington, DC: Author. Retrieved from http://www.ntia.doc.gov/ntiahome/digitaldivide

Negroponte, N. (1995). *Being digital*. New York: Knopf.

Nelson, B. (1999). Grade-A challenges. *Forbes Magazine, 163*(5), March 8, 128.

Neumann, P. G. (1995). *Computer related risks*. Reading, MA: Addison-Wesley.

Newell, S., Scarbrough, H., Swan, J., & Hislop, D. (1998). Intranets and knowledge management: Complex processes and ironic outcomes. *Proceedings of the Thirty-Second Annual Hawaii International Conference on System Sciences*. Retrieved from http://www.computer.org/proceedings/hicss/0001/00011/00011010abs.htm

NUA. (2002). *How many online?* Retrieved from http://www.nua.ie/surveys/how_many_online

Office of Technology Policy (1998). *The new innovators: Global patenting trends in five sectors*. U.S. Department of Commerce, Office of Technology Policy. Retrieved from http://www.technology.gov/Reports.htm.

Olesen, K., & Myers, M. D. (1999). Trying to improve communication and collaboration with information technology: An action research project which failed. *Information Technology & People, 12*(4), 317–332.

Orlikowski, W. (1993). CASE tools as organizational change: Investigating incremental and radical changes in systems development. *Management Information Systems Quarterly, 17*(3), 309–340.

Orlikowski, W., & Barley, S. R. (2001). Technology and institutions: What information systems research and organization studies can learn from each other. *Management Information Systems Quarterly, 25*, 145–165.

Orlikowski, W., & Iacono, S. (2001). Desperately seeking the "IT" in IT research: A call to theorizing the IT artifact. *Information Systems Research, 12*(2), 121–124.

Orlikowski, W., & Robey, D. (1991). Information technology and the structuring of organizations. *Information Systems Research, 2*(2), 143–169.

Patterson, E. S., Cook, R. I., & Render, M. L. (2002). Improving patient safety by identifying side effects from introducing bar coding in medication administration. *Journal of the American Medical Informatics Association, 9*(5), 540–553.

Pew Internet Life Project. (2002). *The digital disconnect: The widening gap between Internet-savvy students and their schools.* August 14, 2002. Retrieved from http://www.pewinternet.org/reports/toc.asp?Report=67

Pew Internet Life Project. (2003a). *America's online pursuits: The changing picture of who's online and what they do.* December 22. Retrieved from http://www.pewinternet.org/reports/toc.asp?Report=106

Pew Internet Life Project. (2003b). *Internet use by region in the United States.* August 27. Retrieved from http://www.pewinternet.org/reports/toc.asp?Report=98

Pew Internet Life Project. (2003c). *The ever-shifting Internet population: A new look at Internet access and the digital divide.* April 16. Retrieved from http://www.pewinternet.org/reports/toc.asp?Report=88

Pew Internet Life Project. (2004). Home page. Retrieved from http://www.pewinternet.org

Pfeffer, J. (1981). *Power in organizations.* Marshfield, MA: Pitman Publishing.

Pickin, J. G. (1996). *The environmental impacts of paper-consuming office technologies in Australia: Executive summary.* Melbourne, Australia: Australian Conservation Foundation. Retrieved from http://www.peg.apc.org/~acfenv/paperrpt.htm

Pino, J., & Mora, H. (1998). Scheduling meetings using participants' preferences. *Information Technology & People, 11*(2), 140–151. Retrieved from http://www.dur.ac.uk/~dcs8s05/journal.htm

Poltrock, S. E., & Grudin, J. (1994). Organizational obstacles to interface design and development: Two participant observer studies. *ACM Transactions on Computer–Human Interaction,* 1(1), 52–60.

Porter, R. W., & Hicks, I. (2000). *Knowledge utilization and the process of policy formation: Towards a framework for Africa.* SARA project (USAID). Retrieved from http://www.aed.org/publications/knowledgeutil.pdf

Poole, M. S., & DeSanctis, G. (1990). Understanding the use of group decision support systems. In J. Fulk & C. Steinfield (Eds.), *Organizations and communication technology* (pp. 173–193). Newbury Park, CA: Sage.

President's Committee of Advisors on Science and Technology: Panel on Educational Technology (PCAST). (1997). *Report to the President on the use of technology to strengthen K-12 education in the United States,* March. Washington, DC: White House. Retrieved from http://www. whitehouse.gov/WH/EOP/OSTP/NSTC/PCAST/k-12ed.html

President's Information Technology Advisory Council (PITAC). (1999). *PITAC report to the President: Information technology: Transforming our society.* Washington, DC: National Coordination Office for Computing, Information and Communication. Retrieved from http://www.ccic. gov/ac/sew-99.pdf

President's Information Technology Advisory Committee. (1998). *Socio-Economic and workforce impacts: Interim report on social, economic, and workforce issues.* Washington, DC: National Coordination Office for Computing, Information and Communication. Retrieved from http:// www.ccic.gov/ac/sew-04Nov98.pdf

Progressive Policy Institute. The New Economy Task Force. (1999). *Rules of the road: Governing principles for the new economy.* Retrieved from http://www.ppionline.org/ppi_ci.cfm?knlgAreaID=107&subsecID= 123&contentID=1268

Progressive Policy Institute. The New Economy Task Force. (1998). *The Internet and Society: Universal Access, Not Universal Service.* Retrieved from http://www.ndol.org/documents/internet_society.pdf

Porter, R., & Hicks I. (2000). *Knowledge utilisation and the process of policy formation: Towards a framework for Africa.* SARA project (USAID).

Qamar, M. K. (1998). *Status and constraints of training of extension staff in Africa: An international view* (posted 3/98). Retrieved from http://www.fao.org/WAICENT/faoinfo/sustdev/EXdirect/EXan0023.htm

Quintas, P. R. (1994). Programmed innovation: Trajectories of change in software development. *Information Technology & People, 7*(1), 25–47.

Rallis, S. F., & MacMullen, M. M. (2000). Inquiry-minded schools: Opening doors for accountability. *Phi Delta Kappan, 81*(9), 766–773. Retrieved from http://www.pdkintl.org/kappan/kral0006.htm

Rawlins, G. J. E. (1996). *Moths to flame: Seductions of computer technology.* Cambridge, MA: MIT Press.

Rawlins, G. J. E. (1997). *Slaves of the machine: The quickening of computer technology.* Cambridge, MA: MIT Press.

Redish, E. F. (1996). New models of physics instruction based on physics education research: Part I. *Proceedings of the Deustchen Physikalischen*

Gesellschaft Jena Conference. Retrieved from http://www.physics.umd.edu/rgroups/ripe/papers/jena/jena.html

Roberts, K. H., & Bea, R. G. (2001). When systems fail. *Organizational Dynamics, 29*(3), 179–191.

Robey, D. (1997). The paradox of transformation: Using contradictory logic to manage the organizational consequences of information technology. In C. Sauer & P. Yetton (Eds.), *Steps to the future: Fresh thinking on the dynamics of organizational transformation.* San Francisco: Jossey-Bass.

Robey, D., & Markus, M. L. (1998). Beyond rigor and relevance: Producing consumable research about information systems. *Information Resources Management Journal, 11*(Winter), 7–15.

Rochlin, G. I. (1997). *Trapped in the Net: The unanticipated consequences of computerization.* Princeton, NJ: Princeton University Press.

Sachs, P. (1995). Transforming work: Collaboration, learning and design. *Communications of the ACM, 38*(9), 36-45.

Saetnan, A. R. (1991). Rigid politics and technological flexibility: The anatomy of a failed hospital innovation. *Science, Technology, & Human Values, 16*(4), 419–447.

Salzman, H. (1989). Computer-aided design: Limitations in automating design and drafting. *IEEE Transactions on Engineering Management, 36*(4), 252–261.

Salzman, H., & Rosenthal. S. (1994). *Software by design: Shaping technology and the workplace.* New York: Oxford University Press.

Sawyer, S. (2001). Information systems development: A market-oriented perspective. *Communications of the ACM, 44*(11), 97–102.

Sawyer, S., & Chen, T. (2002). Conceptualizing information technology and studying information systems: Trends and issues. In M. Myers, E. Whitley, E. Wynn, & J. DeGross (Eds.), *Global and organizational discourse about information technology,* (pp. 109–131). London: Kluwer.

Sawyer, S., Crowston, K., Wigand, R., & Allbritton, M. (2003). The social embeddedness of transactions: Evidence from the residential real estate industry. *The Information Society, 19*(2), 135–155.

Sawyer, S., & Eschenfelder, K. R. (1998). Corporate IT skill needs: A case study of BigCo. *Computer Personnel, 19*(2), 27–41.

Sawyer, S., & Eschenfelder, K. (2002). Social Informatics: Perspectives, examples, and trends. In B. Cronin (Ed.), *Annual Review of Information Science and Technology, 36* (pp. 426–465). Medford, NJ: Information Today Inc./ASIST.

Sawyer, S., Farber, J., & Spillers, R. (1997). Supporting the social processes of software development teams. *Information Technology & People, 10*(1), 46-62.

Sawyer, S., & Rosenbaum, H. (2000). Social Informatics in the information sciences: Current activities and emerging directions. *Informing Science, 3*(2). Retrieved from http://inform.nu/Articles/Vol3/indexv3n2.htm

Sawyer, S., & Tapia, A. (2003). The computerization of work: A Social Informatics perspective. In J. George (Ed.), *Social issues of computing* (pp. 93–109). New York: Oxford.

Scacchi, W. (2004). Socio-technical design. In W. Bainbridge (Ed.), *The encyclopedia of human–computer interaction*. New York: Berkshire.

Schmidt, K., & Bannon, L. J. (1992). Taking CSCW seriously: Supporting articulation work. *Computer Supported Cooperative Work (CSCW): An International Journal.* Vol. 1, n. 1–2, p. 7–40.

Schofield, J. (1995). *Computers and classroom culture.* New York: Cambridge University Press.

Schon, D. A. (1987). *Educating the reflective practitioner.* San Francisco: Jossey Bass.

Schonfeld, E. (1998). Schwab puts it all online: Schwab bet the farm on low-cost Web trading and in the process invented a new kind of brokerage. *Fortune, 138*(11), December 7. Retrieved from http://cgi.pathfinder.com/fortune/technology/1998/12/07/sch.html

Schultze, U., & Leidner, D. E. (2002). Studying knowledge management in information systems research: Discourses and theoretical assumptions. *MIS Quarterly, 26*(3), 213–242.

Scoppio, G. (2000). Common trends of standardisation, accountability, devolution and choice in the educational policies of England, U.K., California, U.S.A., and Ontario, Canada. *Current Issues in Comparative Education, 2*(2), April 30. Retrieved from http://www.tc.columbia.edu/cice/vol02nr2/gsart1.htm

Scott, W. (2001). *Institutions and organizations.* Thousand Oaks, CA: Sage.

Sellen, A., & Harper, R. (1997). Paper as an analytic resource for the design of new technologies. *CHI 97 Electronic Publications: Papers.* Retrieved from http://www.acm.org:82/sigs/sigchi/chi97/proceedings/paper/ajs.htm

Senn, J. (1998). The challenge of relating IS research to practice. *Information Resources Management Journal, 11*(Winter), 23–28.

Silver, M. S., Markus, M. L., & Beath, C. M. (1995). The Information Technology Interaction Model: A foundation for the MBA core course. *Management Information Systems Quarterly, 19*(3), 361–390.

Sneyd, K. P., & Rowley, J. (2003). Linking strategic objectives and operational performance: An action research-based exploration. *Measuring Business Excellence, 8*(3), 42–52.

Sproull, L., & Goodman, P. S. (1989). Technology and organizations: Integration and opportunities. In P. S. Goodman & L. Sproull (Eds.), *Technology and organizations* (pp. 254–266). New York: Jossey-Bass.

Sproull, L., & Kiesler, S. (1991). *Connections*. Cambridge, MA: MIT Press.

Star, S. L., & Ruhleder, K. (1996). Steps toward an ecology of infrastructure: Design and access for large information spaces. *Information Systems Research, 7*, 111–133.

Steering Committee. (1998). *From awareness to action: Integrating ethics and social responsibility across the computer science curriculum*, Project ImpactCS. Arlington, VA: National Science Foundation.

Steering Committee on Research Opportunities Relating to Economic and Social Impacts of Computing and Communications. Computer Science and Telecommunications Board. Commission on Physical Sciences, Mathematics, and Applications. National Research Council. (1998). *Fostering research on the economic and social impacts of information technology: Report of a workshop*. Retrieved from http://www.nap.edu/readingroom/books/esi/ch4.html#43

Stoll, M. (1998). Lessons on laptops. *Christian Science Monitor*, June 9. Retrieved from http://csmonitor.com/cgi-bin/durableredirect.pl?/durable/1998/06/09/Fp5151-csm.htm

Streeter, T. (1996). *Selling the air: A critique of the policy of commercial broadcasting in the United States*. Chicago: University of Chicago Press.

Suchman, L. (1987). *Plans and situated actions: The problem of human–machine communications*. Cambridge, UK: Cambridge University Press.

Suchman, L. (1996). Supporting articulation work. In R. Kling (Ed.), *Computerization and controversy: Value conflicts and social choices, 2nd ed.* (pp. 407–413). San Diego: Academic Press.

Suchman, L. (2002). Figuring service in discourses of ICT: The case of software agents. In E. Wynn, E. Whitley, M. Myers, & J. DeGross (Eds.), *Global and organizational discourse about information technology* (pp. 33–45). New York: Kluwer.

Stix, G. (1996). Aging airways. In R. Kling (Ed.), *Computerization and controversy: Value conflicts and social choices, 2nd ed.* (pp. 793–809). San Diego: Academic Press.

Talbott, S. (1995). *The future does not compute.* Sebastopol, CA: O'Reilly & Associates.

Tavistock Institute. (1998). *Planning and sequencing successful organisational change: An interactive workshop introducing a new methodology.* Retrieved from http://www.tavinstitute.org/workshop

Tenner, E. (1996). *Why things bite back: Technology and the revenge of unintended consequences.* New York: Knopf.

Thorngate, W. (2001). The social psychology of policy analysis. *Journal of Comparative Policy Analysis, 3*(1), 85–113.

Truex, D. P. (2001). Three issues concerning relevance in IS research: Epistemology, audience and method. *Communications of the Association of Information Systems, 6*(24). Retrieved from http://cais.isworld.org/articles/6-24/default.asp?View=html&x=47&y=6

Truex, D., Baskerville, R., & Klein, H. (1999). Growing systems in emergent organizations. *Communications of the ACM, 42*(8), 117–124.

UCLA Center for Communication Policy. (2004). *The 1st UCLA world Internet report.* Retrieved from http://www.ccp.ucla.edu/pages/NewsTopics.asp?Id=45

Ullman, E. (1997). *Close to the machine: Technophilia and its discontents.* San Francisco: City Lights.

University of Arizona. Arizona Cooperative Extension. (2002). *About us.* Retrieved from http://ag.arizona.edu/extension/about/index.html

University of Connecticut. Cooperative Extension System. (n.d.). Welcome to Connecticut Cooperative System. Retrieved from http://www.canr.uconn.edu/ces/welcome.html

Unsworth, J. (1997). Documenting the reinvention of text: The importance of failure. *The Journal of Electronic Publishing, 3*(2). Retrieved from http://www.press.umich.edu/jep/03-02/unsworth.html

U.S. Congress, Office of Technology Assessment (OTA). (1982). *Computerized criminal history systems.* Washington DC: U.S. Government Printing Office.

U.S. Congress, Office of Technology Assessment (OTA). (1990). *Critical connections: Communication for the future.* OTA-CIT-407. Washington, DC: U.S. Government Printing Office. Retrieved from http://www.wws.princeton.edu/~ota/disk2/1990/9014_n.html

U.S. Congress, Office of Technology Assessment (OTA). (1991). *Rural America at the Crossroads: Networking for the Future.* Washington, DC: U.S. Government Printing Office.

U.S. Department of Commerce. (1998). *Update: America's new deficit.* Washington, DC: Office of Technology Policy.

U.S. Department of Commerce. (1999). *The emerging digital economy.* Washington, DC: U.S. Government Printing Office.

U.S. Government, The White House. (1997). *A framework for global electronic commerce,* July 1. Retrieved from http://www.ecommerce. gov/framewrk.htm

Vessey, I., & Sravanapudi, P. (1995). CASE tools as collaborative support technologies. *Communications of the ACM, 38*(1), 83–95.

Virtual Society?. (2000). *An overview of the Virtual Society? Programme.* Retrieved from http://www.brunel.ac.uk/research/virtsoc/over.htm

Wagner, I. (1993). A web of fuzzy problems: Confronting the ethical issues. *Communications of the ACM, 36*(4), 94–101.

Walsham, G. (1998). IT and changing professional identity: Micro studies and macro-theory. *Journal of the American Society for Information Science, 49*(12), 1081–1089.

Webster, F. (1995). *Theories of the information society.* New York: Routledge.

Weingarten, F. (1996). Technological change and the evolution of information policy. *American Libraries, 27*(11), 45–47.

Weingarten, F., & Overbey, P. (1995). *Culture, society and advanced information.* Computing Research Association. Retrieved from http://www.cra.org/Policy/reports/aspects

Wellman, B. (2001). Computer networks as social networks. *Science, 293*(14), 2031–2034.

Wellman, B., & Haythornthwaite, C. (Eds.). (2002). *Internet and everyday life.* London: Blackwell Publishing.

Wegner, E., McDermott, R., & Snyder, W. M. (2002). *Cultivating communities of practice.* Cambridge, MA: Harvard Business School Press.

Westfall, R. (1999). An IS research relevance manifesto. *Communications of the Association of Information Systems, 2*(24). Retrieved from http://cais.isworld.org/articles/2-14/default.asp?View=html&x=65&y=11

Wildavsky, A. (1979). *Speaking truth to power: The art and craft of policy analysis.* Boston: Little, Brown and Co.

Williams, R. (1997). The social shaping of information and communications technologies. In H. Kubicek, W. Dutton, & R. Williams (Eds.), *The social shaping of information superhighways: European and American roads to the information society* (pp. 299–338). New York: St. Martin's.

Williams, R., & Edge, D. (1996). The social shaping of technology. *Research Policy, 25*, 856–899.

Wilson, M., & Howcroft, D. (2002). Re-conceptualising failure: Social shaping meets IS research. *European Journal of Information Systems, 11*, 236–250.

Wilson, S., Bekker, M., Johnson, P., & Johnson, H. (1997). Helping and hindering user involvement: A tale of everyday design. *CHI '97 Electronic Publications: Papers*. Retrieved from http://www1.acm.org:81/sigchi/chi97/proceedings/paper/sw-obf.htm

Wolowitz, S., & Diana, A. J. (1998). Unexpected SEC issues are arising online: Particular features of the Internet can affect insider trading, stock manipulation and other areas subject to SEC regulation. *The National Law Journal*, February 9, B07.

Woods, D. (2001). *Human-centered design of automated agents and human-automation team play*. Columbus, OH: Cognitive-Systems Engineering Laboratory, Ohio State University.

Woods, D., & Patterson, E. (2001). How unexpected events produce an escalation of cognitive and coordinative demands. In P. Hancock & P. Desmond (Eds.), *Stress, workload, and fatigue* (pp. 290–302). Mahwah, NJ: Lawrence Erlbaum.

Xenakis, J. J. (1996). Taming SAP. *CFO: The Magazine for Senior Financial Executives, 12*(3), 23–30.

Glossary

Analytical orientation: An approach to research on ICTs in organizational and social contexts that focuses on the institutional nature of ICTs. Also, empirical studies that are organized to contribute to such theorizing.

Articulation: The additional work needed to make ICTs useful. These often take the form of work-arounds, learning effects, manual responses to "clumsy" automation, and the extra work to keep systems running.

Artifact: An object created by people for practical use.

Attribution theory: From the field of psychology, this theory describes the perception of people or things as causal agents. Attribution theory has three steps: perception, judgment, and attribution. For example, if a person witnesses someone throwing trash out of a car window, that person might think the following: I saw you litter (Perception of action); You wanted to litter (Judgment of intention); You are a disgusting person (Attribution of disposition).

CIM system: Computer integrated manufacturing system. A system that utilizes a shared manufacturing database for engineering design, manufacturing engineering, factory production, and information management. See also *Enterprise integration system.*

Communication system: A system whose primary focus is to transmit messages (text, image, and voice). Increasingly these are digital transmissions.

Configuration: The complex array of standardized and customized computer and ICT elements. This can mean the personalization of browser, interface, and office suite software for a particular user on a particular machine, or the set of processing links, data passing rules, and collection of artifacts that are linked together into a multi-vendor and multi-platform system. Configurations are dynamic, local, and connect multiple levels of artifacts, machines, processes, and people.

Configurational technology: Technology, such as a computer integrated manufacturing system, that is configured to meet the specific structure, working methods, and requirements of an organization. An intranet is an example of a configurational technology.

Contract development: Within the context of system design, the user organization is known from the outset and the development organization is identified after the contract is awarded.

Critical orientation: The examination of ICTs from perspectives that do not automatically and "uncritically" adopt the goals and beliefs of the groups that commission, design, or implement specific ICTs.

Custom development: See *In-house development*.

DBMS: Database management system. Computer programs that facilitate the retrieval, modification, and storage of information in a database. A DBMS can run on a PC or a mainframe.

Digital library: A federated repository of documents—encoded in multimedia and/or digital formats—that are locally or remotely accessible via computer networks.

Direct effect theories: Theories that expect or anticipate that the social and organizational consequences of computerization are linear or simple. For example, access to the Internet will improve education, computers make people productive, the Internet will make information freely available to everyone, and enterprise integration systems make organizations more efficient.

Displacement: The replacement of workers by technology.

Distance education: The delivery of a course, workshop, or degree via electronic means to participants or students located away from the delivery origination point in a synchronous or asynchronous format.

Domain: A sphere of work or activity.

Domain of action: A sphere of work or activity in which a particular community of people is engaged.

Electronic commerce: The conducting of business or the facilitation of commerce on the Internet.

Electronic publishing: The creation, publication, and dissemination of digital documents on the Internet.

Empirically anchored theory: Theories that are driven by extensive evidence. Empirically anchored theories are often situated in particular domains and allow for variations in outcomes. Such theories help analysts and scholars to better anticipate contradictory consequences, explain successes and failures, and predict conditions under which [ICT] systems will fail by some criteria.

End-user: Anyone who uses a computer. Same as *user*. It is a conceptual error to think that there exists a generic user. From a Social Informatics perspective, the term *user* is a convenient fiction that marginalizes or homogenizes the individual.

Enterprise integration system: A system that integrates information across parts or all of an enterprise. Information integration essentially consists of providing the right information, at the right place, at the right time for an enterprise operation. These are also known as enterprise resource packages (ERP) and enterprise systems (ES). See also *CIM system*.

Firewall: A system designed to prevent unauthorized access to or from a private network such as an Intranet.

GroupWare: Computer software and hardware that allows people to work together locally or remotely.

Host: A computer system that holds data and is accessed by remote users using TCP/IP.

Human-centered system: A system designed to be used easily and effectively by people such that the global needs of the organization served by the system are met. See also *Organization-centered system*.

Human–computer interaction: Often known as HCI, this is the discipline that looks at the design, evaluation, and implementation of interactive computing systems for human use. Human–computer interface (also known as HCI) is a subset of this discipline that focuses on the design of the machine's interface (the portion of the system that people see or use).

ICT: Information and communication technology—artifacts and practices for recording, organizing, storing, manipulating, and communicating information. ICTs include a wide array of artifacts such as telephones, faxes, photocopiers, movies, books, and journal articles. They also include practices such as software testing methods and approaches to cataloging and indexing documents in a library.

ICT-oriented curricula: Courses whose content is focused on providing students with theories, models, techniques, and tools to enable the design, development, implementation, and support of ICTs.

ICT-oriented education: The various disciplines (e.g., information science, information systems, computer science, etc.) that educate students in the design, development, implementation, and support of ICTs.

ICT-oriented student: A student who takes courses or conducts research within the framework of ICT-oriented education.

In-house development: Within the context of system design, both the eventual users and the developers are known at the project outset. This is also known as *custom development*, where a specific external developer is engaged from the start in producing or configuring a system for a specific customer.

Information processing system: A system that performs operations upon data.

Intensification: An increase in work pressure and/or working hours due to the presence and/or use of computers.

Internet: A decentralized network based on TCP/IP protocols. It serves as a digital and nearly global communications network. Unlike online services and intranets, which are centrally controlled, the Internet is decentralized by design. Each Internet computer, called a host, is independent. Its operators can choose which Internet services to use and which local services to make available to the global Internet community.

Intranet: A network based on TCP/IP protocols belonging to an organization, usually a corporation, accessible only by the organization's members, employees, or others with authorization. An intranet's Web site looks and acts just like any other Web site, but the firewall surrounding an intranet fends off unauthorized access. Like the Internet itself, intranets are used to share information. Secure intranets are now the fastest-growing segment of the Internet because they are much less expensive to build and manage than private networks based on proprietary protocols.

Mainframe: A very large computer (or supercomputer) that is capable of supporting hundreds of users and running numerous programs simultaneously.

MRP: Material Requirements Planning systems. MRPs—or computerized inventory control and production scheduling systems—are transaction-oriented ICTs whose data refers to material. The purchasing departments of manufacturing firms rely on MRPs as a means of reducing inventories, thus reducing costs.

Normative orientation: Research whose aim is to recommend alternatives for professionals who design, implement, use, or make policy about ICTs. It has the explicit goal of influencing practice by providing empirical evidence illustrating the varied outcomes that occur as people work with ICTs in a wide range of organizational and social contexts.

Operations research: The mathematical science that seeks to carry out complicated operations with maximum efficiency.

Organization: An administrative or functional structure and the personnel associated with that structure.

Organization-centered system: A system designed to meet the needs of an organization and not its personnel. This system makes excessive demands upon people in order to use it effectively. See also *Human-centered system*.

Organizational Informatics: Social Informatics analyses that are bounded within organizations, where the primary participants are located within a few identifiable organizations. Many studies of the roles of computerization in shaping work and organizational structures fit within Organizational Informatics. See also Social Informatics.

Organizational politics: Exercises of power or influence within an organization in order to further individual or group self-interests and personal goals or to strike a balance between competing interests.

PC: Personal computer. Based on the Intel microprocessor or Intel-compatible microprocessor, a PC is designed to run a DOS, Windows, or LINUX (a version of UNIX designed for PCs) operating system.

Personal digital assistant: Also known as PDA. A handheld ICT device that often combines computing, telephone, and fax capabilities. Instead of using a keyboard for the inputting of data, a PDA uses a stylus that can incorporate handwriting recognition features. Voice recognition is another feature incorporated in some PDAs.

Product development: Within the context of system design, the developers are known from the outset, but the users typically remain unknown until the product is marketed.

Productivity paradox: Attributed to Nobel Laureate Robert Solow, this term addresses the phenomenon of computers not making an impact on productivity statistics in spite of their widespread use.

Social determinism: The theory that posits that society shapes technology. It is the opposite of *technological determinism*.

Social Informatics: The interdisciplinary study of the design, uses, and consequences of ICTs that takes into account their interactions with institutional and cultural contexts.

Socio-technical package: A way of viewing an ICT as a complex, interdependent system comprised of people, computer hardware and software, techniques, and data.

System: A network composed of discrete components that distributes information electronically.

Systems rationalism: A perspective that conceptualizes ICTs as rule-bound and carefully structured and then generalizes these characteristics to people, groups, and organizations.

TCP/IP protocols: Transmission control protocol/Internet protocol. UNIX-based communication protocols that are used to connect hosts on the Internet in order to transmit data.

Technological determinism: The theory that posits that technology shapes society. It is the opposite of *social determinism*.

Telecommuting: An alternative method of working wherein an individual, employed by an organization, performs his or her job away from the organization [typically at home] through the use of ICTs.

UNIX: A multi-user, multitasking operating system developed by Bell Laboratories in the early 1970s. UNIX is the standard operating system of workstations.

Usability testing: The testing of a system by its developers and its intended users to determine if that system meets the requirements and abilities of those users. Used during the development of computer interfaces to determine their efficacy and reliability.

User: See *End-user*.

User-friendly: A system designed so that it can be utilized by as broad a range of individuals as possible.

Reviews and Anthologies of Social Informatics Research

Berleur, Jacques, Andrew Clement, Richard Sizer, & Diane Whitehouse (Eds.). (1991). *The information society: Evolving landscapes.* New York: Springer Verlag.

Bishop, Ann, & Leigh Star. (1996). Social informatics for digital libraries. In *Annual Review of Information Science and Technology,* 31 (pp. 301–403). Washington, DC: ASIST.

Bowker, Geoffrey, Susan Leigh Star, William Turner, & Les Gasser (Eds.). (1997). *Social science, technical networks and cooperative work: Beyond the great divide.* Hillsdale, NJ: Erlbaum.

Commission on Physical Sciences, Mathematics, and Applications. (1998). *Fostering research on the economic and social impacts of information technology.* Washington, DC: National Academy Press.

Dutton, William (Ed.). (1997). *Information & communication technologies: Vision & realities.* Oxford, UK: Oxford University Press.

Dutton, William H. (1999). *Society on the line: Information politics in the digital age.* Oxford and New York: Oxford University Press.

Garson, G. David. (2000). *Social dimensions of information technology: Issues for the new millennium.* Hershey, PA: IDEA Publishing Group.

Huff, Chuck, & Thomas Finholt (Eds.). (*1994). Social issues in computing: Putting computing in its place.* New York: McGraw-Hill.

Jones, Steve. (2000). *Virtual culture: Identity and communication in cybersociety.* Thousand Oaks, CA: Sage.

Kiesler, Sara (Ed.). (1997). *The culture of the internet.* Mahwah, NJ: Erlbaum.

Kling, Rob. (1996). *Computerization and controversy: Value conflicts and social choices. 2nd ed.* San Diego: Academic Press.

Kling, Rob, & Tom Jewett. (1994). "The Social Design of Worklife With Computers and Networks: An Open Natural Systems Perspective." In Rob Kling & Tom Jewett (Eds.), *Advances in Computers,* 39. New York: Academic Press.

Kubicek, Herbert, William Dutton, & Robin Williams (Eds.). (1997). *The social shaping of information highways: European and American roads to the information society.* New York: Campus Verlag/St. Martins's.

Lievrouw, L., & S. Livingstone. (2002). *The handbook of new media*. London: Sage.

Mansell, Robin, & Roger Silverstone. (1995). *Communication by design: The politics of information and communication technologies*. Oxford: Oxford University Press.

Sawyer, S., & K. Eschenfelder. (2002). Social informatics: Perspectives, examples and trends. In B. Cronin (Ed.), *Annual Review of Information Science and Technology, 36* (pp. 427–465). Medford, NJ: Information Today/ASIST.

Smith, Marc, & Peter Kollock (Eds.). (1998). *Communities in cyberspace*. London: Routledge.

Wellman, Barry, J. Salaff, D. Dimitrova, L. Garton, M. Gulia, & C. Haythornthwaite. (1996). Computer networks as social networks: Collaborative work, telework, and virtual community. *Annual Review of Sociology, 22*, 213–238.

Wellman, B. (Ed.). (1999). *Networks in the global village*. Boulder, CO: Westview Press.

Structure and Process of the 1997 Social Informatics Workshop

A workshop, funded in part by the National Science Foundation, on *Advances in Social informatics* was held at Indiana University in Bloomington, Indiana (US), in November 1997. The main focus was the articulation of two emerging interdisciplinary research domains, called *Organizational Informatics* and *Social Informatics*. These names represent the research—and researchers—that focus on the social dimensions of the integration of computerization and networked information into social and organizational life.

The main purpose of this workshop was to clarify systematically the domains of Social Informatics by exploring the state of knowledge about the integration of computerization and networked information into social and organizational life, and the roles of information and communication technology (ICT) in social and organizational change. This was done by gathering a group of experienced and recognized scholars and researchers whose work overlaps with the concerns of Social Informatics (a list of participants is included in Appendix C). By focusing on the range of concepts, theories, and findings that are being brought to bear in the study of the roles of ICTs in organizational and social change, participants in the workshop assessed the fields of organizational informatics and Social Informatics, defined their research frontiers, and reflected on their place in the academic landscape.

The workshop brought together twenty-five scholars and researchers from several different disciplines, including information science, sociology, communications, information systems, social psychology, computer science, and anthropology, for two days of intensive interaction, the goals of which were to:

- Provide input into this report, which more sharply characterizes organizational and social informatics.

- Better identify the state of knowledge.

- Outline a set of research issues that could constitute effective and plausible advances in Social and Organizational Informatics.

Workshop participants were invited based on two criteria. First, their research interests overlapped with the concerns of Social Informatics. Their work included research on the social contexts of ICT design, implementation, and use in a variety of social and organizational settings. Second, they had interdisciplinary experience and a willingness to engage in dialog with researchers working in different but cognate disciplines. A workshop computer conference was established and participants were able to meet and interact with each other through this conference, providing introductions and statements of their research interests. The online discussion was seeded by the workshop organizers, who used the conference listserv to disseminate the workshop agenda and a list of discussion questions. In addition, a workshop Web site was established and used to gather together and make available documents considered by the organizers, the advisory committee, and participants to be within the domain of Social Informatics.

At the workshop, participants discussed the range of research directions and problems in social informatics that were considered strategic, timely, and tractable given the present state-of-the-field. They discussed research directions based on a range of criteria, such as the extent to which the research would encourage the progress of Social Informatics; the likelihood of success, both in the short and long run; current and future funding possibilities; and national needs. Participants worked in teams to prepare short position papers on the definition and domain of Social Informatics, potential research problems, and directions. There were also sessions devoted to discussing strategies for expanding the research community interested in Social Informatics and strategies for incorporating the concerns of Social Informatics into graduate and undergraduate curricula.

Triangle diagram[1]

Workshop participants developed the triangle diagram on page 193 to help portray the scope of Social Informatics.

The diagram and definition of Social Informatics help to emphasize the key idea that ICTs do not exist in social or technological isolation. Their "cultural and institutional contexts" influence the ways they are developed, the kinds of workable configurations that are proposed, how they are implemented and used, and the range of consequences for organizations and other social groupings.

Endnote

1. Based on *The Culture of Technology* by Arnold Pacey (Cambridge, MA: MIT Press, 1985).

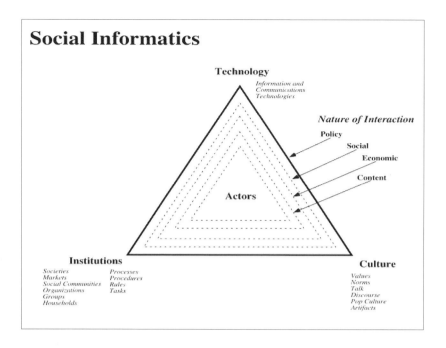

Social Informatics

Technology
Information and
Communications
Technologies

Nature of Interaction

Policy

Social

Economic

Content

Actors

Institutions

Societies   *Processes*
Markets   *Procedures*
Social Communities *Rules*
Organizations   *Tasks*
Groups
Households

Culture

Values
Norms
Talk
Discourse
Pop Culture
Artifacts

1997 Social Informatics Workshop Participants

Mark Ackerman	C. Suzanne Iacono	Steve Sawyer (#)
Chrisianthi Avgerou	Rob Kling (#)	Charles Steinfield
Gerry Bernbom	Ken Kraemer	Lucy Suchman
Werner Beuschel	Roberta Lamb	Nancy Van House
Andrew Clement	Leah Lievrouw	Alladi Venkatesh
Holly Crawford (#)	Chris Ogan	Rick Weingarten
Andrew Dillon	Sheizaf Rafaeli	Susie Weisband (#)
Linda Garcia	Ron Rice	Barry Wellman
Les Gasser	Howard Rosenbaum (#)	Bob Zmud
Carol Hert	Rohan Samarajiva	

(#) Co-author of the workshop report provided to the NSF

APPENDIX D

Additional Reviewers

We gratefully acknowledge these scholars for their comments and insights. Their interest and attention have substantially improved this book.

Phil Agre	Elisabeth Davenport	Carleen Maitland
Steve Alter	Jonathan Grudin	Robin Mansell
Geoff Bowker	Rudi Hirschheim	Robin Peek
Laku Chidambarum	Anne Hoag	Sharon Ross
Lisa Covi	Andrea Hoplight Tapia	Ralf Shaw
Blaise Cronin	Chuck Huff	Leigh Star
Mary Culnan	Lynette Kvasny	Rolf Wigand

About the Authors

Rob Kling grew up in northern New Jersey. He completed his undergraduate studies at Columbia University (1965) and his graduate studies, specializing in Artificial Intelligence, at Stanford University (1967, 1971). Between 1966 and 1971 he held a research appointment in the Artificial Intelligence Center at the Stanford Research Institute. He held his first professorship in Computer Science at the University of Wisconsin-Madison from 1970 to 1973. He was on the faculty of the University of California at Irvine from 1973–1996 and held professorial appointments at UCI's Center for Research on Information Technology and Organizations and Graduate School of Management. In August 1996, he moved to Indiana University, Bloomington as Professor of Information Science and Information Systems. At Indiana he directed the Master of Information Science degree program. At Indiana, Dr. Kling also founded and directed the Center for Social Informatics. This interdisciplinary research center became the intellectual center of social informatics and was renamed the Rob Kling Center for Social Informatics. The center is run jointly by Indiana's Graduate School of Library and Information Science and the School of Informatics.

From the early 1970s Dr. Kling studied the social opportunities and dilemmas of computerization for managers, professionals, workers, and the public. Dr. Kling examined computerization as a social process with technical elements. He studied how intensive computerization transforms work and how computerization entails many social choices. He also studied the ways that complex information systems and expert systems are integrated into the social life of organizations. He conducted research in numerous kinds of organizations, including local governments, insurance companies, pharmaceutical firms, and hi-tech manufacturing firms. He wrote about the value conflicts implicit in and social consequences of computerization, which directly affects the public. He was studying the effective use of electronic media to support scholarly and professional communication when he passed away in May 2003. Some of that work was posthumously published in a volume he co-edited, *Designing Virtual Communities in the Service of Learning* (Cambridge University Press, 2004).

Dr. Kling was a co-author of *Computers and Politics: High Technology in American Local Governments* (Columbia University Press, 1982), which examined how computerization reinforces the power of already powerful groups. He was co-editor of *PostSuburban California: The Transformation of Postwar Orange County* (University of California Press, 1990), which examined the ways that Orange County, California is organized in a new social form beyond the traditional city and suburb, one that is spatially decentralized, functionally specialized, and mixes a rich array of residences, commerce, industry, services, government, and the arts. *PostSuburban California* won the Thomas Athearn Award from the Western Historical Society in 1992 and was reissued in paperback in 1995. *Computerization and Controversy: Value Conflicts & Social Choices* (Academic Press, 1991) examined the social controversies about computerization in organizations and social life, regarding productivity, worklife, personal privacy, risks of computer systems, and computer ethics. In 1996, Dr. Kling was the sole editor of a substantially rewritten 2nd edition of *Computerization and Controversy.*

In addition, Dr. Kling's research was published in over 85 journal articles and book chapters. He presented numerous conference papers, gave invited lectures at many major universities and the National Academy of Sciences, and gave keynote and plenary talks at conferences in the United States, Canada, and Western Europe. He consulted for private firms, nonprofit organizations, the Congress of the United States, the President's Information Technology Advisory Council (under President Clinton), and two foreign governments about the opportunities and problems of computerization. In the late 1990s, he served on the Executive Committee of the U.S. ACM Committee for Computers and Public Policy, the American Sociological Association's Committee on Electronic Publishing, and the AAAS's National Conference of Lawyers and Scientists.

At Irvine, Dr. Kling had been co-director of their doctoral concentration on Computing, Organizations, Policy, and Society. He was Editor- in-Chief of *The Information Society* and served on the editorial and advisory boards of several other scholarly and professional journals including *European Journal of CSCW, Information Technology and People, Social Science Computer Review,* and *Accounting, Management and Information Technology.* He also organized special workshops about the social and managerial aspects of computerization, served on the program committees of several major national conferences, and was chair of an (IFIP) international working group on the Social Accountability of Computing.

Dr. Kling was a visiting Professor at the Copenhagen School of Business and Economics and at the Solvay School of Business at the University of Brussels. He was also a Research Fellow at Harvard University's Program on Information Resources Policy and a Visiting Researcher at the Gessellschaft fur Mathematik und Datenverarbeitung in Bonn, Germany.

Dr. Kling's scholarly and professional accomplishments have been recognized nationally and internationally. In 2002 he was named as a Fellow of the Association of Information Systems. In 2001, he was elected as a Fellow of the American Association for the Advancement of Science. In 1987, he was awarded an Honorary Doctorate in Social Sciences by the Free University of Brussels. In 1983, he received a Silver Core Award from the International

Federation of Information Processing Societies. In 1984, he received a Service Award from the Association for Computing Machinery. Following his death, the Association of Information Systems named him the 2004 Leo Award winner for his substantial contributions to information systems research (including the best article award for his co-authored work, "Reconceptualizing Users as Social Actors in Information Systems Research" published in the *MIS Quarterly* in June 2003).

Dr. Howard Rosenbaum is an Associate Professor of Information Science in the School of Library and Information Science at Indiana University and has been on the faculty since 1993. He has an interest in Social Informatics and researches the development of the Internet and its implications for the information professions, electronic business, community networking, and managers and their uses of information in organizations. Recently, Dr. Rosenbaum has been studying trust in electronic business and is working on a study of mobile commerce in Japan. He has led seminars on Electronic Commerce at Napier University Business School in Edinburgh, Scotland, the University of Bath, and the University of Greenwich, in the U.K.

Dr. Rosenbaum has presented his work at the Association for Information Systems, the American Society for Information Science, the Association of Internet Researchers, The HCI International, the International Communications Association, the Canadian Association for Information Science, the American Sociological Association, and other organizations. He is a Fellow in the Center for Social Informatics at Indiana University and the Center for Digital Commerce at Syracuse University.

He has had extensive experience using qualitative methods in a variety of settings to investigate a range of research problems in library and information science. Dr. Rosenbaum teaches classes on electronic business and digital entrepreneurship, information architecture, information systems design, and intellectual freedom, and offers continuing education workshops for information professionals in XML, CSS, and Web page design.

Dr. Rosenbaum has a history of leadership in technological innovation in education. Beginning in 1997, he received the School's Teaching Excellence Recognition Award for three consecutive years and the School's 2001 Trustees Teaching Award. He was named an Ameritech/SBC Fellow in 2000 and received a 2002 IHETS "Indiana Partnership for Statewide Education Award for Innovation in Teaching with Technology." He received a 2003 "MIRA Award for Technological Innovation in Education" and the 2005 Frederic Bachman Lieber Award for Distinguished Teaching. He has also received a 2005 Lilly Foundation grant to develop courses in digital entrepreneurship. Since summer 2004, he has been the Director of the Masters of Information Science program, a leadership role in the school.

Steve Sawyer has degrees from the United States Merchant Marine Academy (1982), the University of Rhode Island (1986) and Boston University (1989, 1995). From 1994 to 1999, Steve served on the faculty of the School of Information Studies at Syracuse University. In 1999 he became a founding member and Associate Professor with the School of Information Sciences and Technology at the Pennsylvania State University at University Park. At Penn State, Steve holds affiliate appointments in Management and Organization (in the Smeal College of Business), Labor Studies and Industrial Relations (in the college of the Liberal Arts), and the Science Technology and Society Program (in the college of Engineering). At Penn State, Steve helped to start the Center for the Information Society. He is also honored to maintain an affiliate appointment with Syracuse's School of Information Studies.

Steve has published more than 40 papers on the roles of information and communications technologies in altering work and its organization. This includes work on software development, the take up and uses of enterprise systems, the uses of Internet and mobile technologies by real estate agents, and the roles of mobile technologies in public safety. This is his first book.

Along with his research, Steve has a passion for learning innovations. While at Syracuse he was named Professor of the Year in 1997. Since coming to Penn State, Steve has been named the first IST Faculty Member of the Year (in 2001) and the school's inaugural George McMurtry award winner for Teaching and Learning Innovation (in 2002).

Sawyer is currently an associate editor for *The Information Society.* He is on the editorial boards of *Information Technology & People,* the *Journal of Information Technology* and the *International Journal of Advanced Media and Communication.* Steve is a member of the Academy of Management, the American Sociological Association, the Association of Computing Machinery (ACM), the Association for Information Systems (AIS), Computer Professionals for Social Responsibility (CPSR), INFORMS, the Institute for Electrical and Electronics Engineers (IEEE), and the International Federation of Information Processing's working group on information systems in organization and society (IFIP WG 8.2), where he is actively involved in its working conferences.

NAME INDEX

A

Abraham, N., 84
Abrams, M., 53
Abran, A., 87
Ackerman, M., 37, 87
Adamson, R. J., 60
Addison, T., 107, 110
Agre, P., 7, 49, 89, 114, 151
Allbritton, M., 20, 29
Allen, J., 94
Allen, W., 27
Ashenhurst, R. L., 89
Aspray, W., 83
Attewell, P., 116, 117, 123
Avgerou, C., 146, 151, 152, 154
Avison, D., 87, 146
Axline, S., 23, 29

B

Bangemann, M., 78
Bannon, L. J., 60
Barley, S. R., 21, 146
Baskerville, R., 37, 45
Bauer, J., 51, 52
Bea, R. G., 110
Beath, C., 87, 96
Bekker, M., 41
Belanger, Y., 56
Bell, D., 77
Benjamin, R. I., 28, 105, 106, 110, 142
Bimber, B., 62
Bowker, G., 49
Braa, J., 90
Brooks, F. P. J., 28, 39, 86
Brown, J. S., 41, 94
Burkhardt, M., 28
Bush, V., 97
Butler, T., 151

C

Cahoon, J., 84
Carmel, E., 96
Carr, A., 85
Carr, N., 106, 107, 110
Checkland, P., 87
Chen, T., 26
Choo, C. W., 121
Christie, J., 57
Clement, A., 24–25, 41
Clinton, W. J., 71, 78
Coley, R. J., 58
Cook, R. I., 35
Cooke, C., 71
Cooper, M. N., 65
Cooprider, J., 26, 28, 41
Copeland, D., 46
Couger, J. D., 89
Cradler, J., 58
Cranor, L. F., 52
Cronin, B., 134
Crowston, K., 20, 29

D

Dalcher, D., 106
Danziger, J. N., 23, 30
Dargan, P. A., 38, 39
Davenport, T. H., 21, 35, 36, 38, 96, 100, 113, 114, 147
Davis, G. B., 89
Denning, P. J., 38, 39
Dertouzous, M., 97
Donaldson, J., 110
Donnermeyer, J. F., 65
Downs, A., 30
Drevin, L., 106
Drummond, H., 105
Duguid, P., 41, 94

More Titles of Interest from Information Today, Inc.

Theories of Information Behavior

Edited by Karen E. Fisher, Sanda Erdelez, and Lynne (E. F.) McKechnie

Here are authoritative overviews of more than 70 conceptual frameworks for understanding how people seek, manage, share, and use information in different contexts. Covering both established and newly proposed theories of information behavior, the book includes contributions from 85 scholars from 10 countries. Theory descriptions cover origins, propositions, methodological implications, usage, and links to related theories.

456 pp/hardbound/ISBN 1-57387-230-X

ASIST Members $39.60 • Nonmembers $49.50

Covert and Overt

Recollecting and Connecting Intelligence Service and Information Science

Edited by Robert V. Williams and Ben-Ami Lipetz

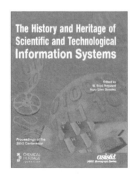

This book explores the historical relationships between covert intelligence work and information/computer science. It first examines the pivotal strides to utilize technology to gather and disseminate government/military intelligence during WWII. Next, it traces the evolution of the relationship between spymasters, computers, and systems developers through the years of the Cold War.

276 pp/hardbound/ISBN 1-57387-234-2

ASIST Members $39.60 • Non-Members $49.50

The History and Heritage of Scientific and Technological Information Systems

Edited by W. Boyd Rayward and Mary Ellen Bowden

The second conference on the history of information science systems covered scientific and technical information systems in the period from WWII through the early 1990s. These proceedings present the papers of historians, information professionals, and scientists on a range of topics including informatics in chemistry, biology and medicine, and information developments in multinational, industrial, and military settings.

440 pp/softbound/ISBN 1-57387-229-6

ASIST Members $36.40 • Non-Members $45.50

Information Representation and Retrieval in the Digital Age

Heting Chu

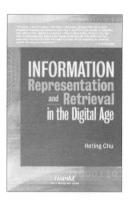

This is the first book to offer a clear, comprehensive view of Information Representation and Retrieval (IRR). With an emphasis on principles and fundamentals, the author first reviews key concepts and major developmental stages of the field, then systematically examines information representation methods, IRR languages, retrieval techniques and models, and Internet retrieval systems.

250 pp/hardbound/ISBN 1-57387-172-9

ASIST Members $35.60 • Nonmembers $44.50

Statistical Methods for the Information Professional

Liwen Vaughan

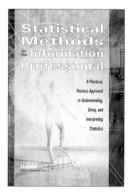

Author and educator Liwen Vaughan clearly explains the statistical methods used in information science research, focusing on basic logic rather than mathematical intricacies. Her emphasis is on the meaning of statistics, when and how to apply them, and how to interpret the results of statistical analysis. Through the use of real-world examples, she shows how statistics can be used to improve services, make better decisions, and conduct more effective research.

240 pp/hardbound/ISBN 1-57387-110-9

ASIST Members $31.60 • Non-Members $39.50

Intelligent Technologies in Library and Information Service Applications

F.W. Lancaster and Amy Warner

In this carefully researched monograph, authors Lancaster and Warner report on the applications of AI technologies in library and information services, assessing their effectiveness, reviewing the relevant literature, and offering a clear-eyed forecast of future use and impact.

214 pp/hardbound/ISBN 1-57387-103-6

ASIST Members $31.60 • Non-Members $39.50

ASIS&T Thesaurus of Information Science, Technology, and Librarianship, Third Edition

Edited by Alice Redmond-Neal and Marjorie M. K. Hlava

The *ASIST Thesaurus* is the authoritative reference to the terminology of information science, technology, and librarianship. This updated third edition is an essential resource for indexers, researchers, scholars, students, and practitioners. An optional CD-ROM includes the complete contents of the print thesaurus along with Data Harmony's Thesaurus Master software. In addition to powerful search and display features, the CD-ROM allows users to add, change, and delete terms, and to learn the basics of thesaurus construction while exploring the vocabulary of library and information science and technology.

Book with CD-ROM: 272pp/softbound/ISBN 1-57387-244-X
$63.95 ASIST members/$79.95 Nonmembers

Book only: 272pp/softbound/ISBN 1-57387-243-1
$39.95 ASIST members/$49.95 Nonmembers

ARIST 40: Annual Review of Information Science and Technology

ARIST, published annually since 1966, surveys the landscape of information science and technology, providing an analytical, authoritative, and accessible overview of recent trends and significant developments. Editor Blaise Cronin is selectively expanding *ARIST*'s footprint in an effort to connect information science more tightly with cognate academic and professional communities.

704 pp/hardbound/ISBN 1-57387-209-1

ASIST Members $79.95 • Non-Members $99.95

Proceedings of the 67th Annual Meeting of the American Society of Information Science & Technology

ASIS&T 2004—Managing and Enhancing Information: Cultures and Conflicts addresses the increasing tension between forces that encourage and discourage integration and cooperation in the global information society. Experts from a range of professional and academic settings around the world share their ideas and successes in nearly 200 contributions.

646 pp/softbound/ISBN 1-57387-222-9

ASIST Members $55.90 • Non-Members $69.90

Best Technology Practices in Higher Education

Edited by Les Lloyd

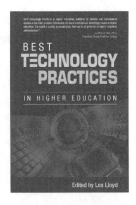

A handful of progressive teachers and administrators are integrating technology in new and creative ways at their colleges and universities, raising the bar for all schools. In his latest book, editor Les Lloyd (*Teaching with Technology*) has sought out the most innovative and practical examples in a range of key application areas, bringing together more than 30 technology leaders to share their success stories. The book's 18 chapters include firsthand accounts of school technology projects that have transformed classrooms, services, and administrative operations. The four major sections are "Best Practices in Teaching and Course Delivery," "Best Practices in Administrative Operations," "Technical or Integrative Best Practices," and "Future Best Practices."

256 pp/hardbound/ISBN 1-57387-208-3 • $39.50

The Successful Academic Librarian
Winning Strategies from Library Leaders

Edited by Gwen Meyer Gregory

The role of academic librarian is far from cut-and-dried. For starters, there are the numerous job classifications: staff or professional employment, full faculty status, various forms of tenure, continuing contract, and/or promotion through academic ranks. While every academic librarian works to meet the research needs of faculty and students, many are expected to assume other obligations as part of a faculty or tenure system— including publication, research, service, and professional activities. If this were not enough to test a librarian's mettle, the widely varying academic focuses and cultures of college and university libraries almost certainly will. This book, edited by academic librarian, writer, and speaker Gwen Meyer Gregory, is an antidote to the stress and burnout that almost every academic librarian experiences at one time or another. Here, Gregory and nearly 20 of her peers take a practical approach to a full range of critical topics facing the profession.

250 pp/hardbound/ISBN 1-57387-172-9

ASIST Members $35.60 • Nonmembers $44.50

To order or for a complete catalog, contact:

Information Today, Inc.

143 Old Marlton Pike, Medford, NJ 08055 • 609/654-6266
email: custserv@infotoday.com • Web site: www.infotoday.com